CICHLIDS
North & of Central America

Donald Conkel

Photos by the author unless specifically credited otherwise

yearBOOKS, INC.
Dr. Herbert R. Axelrod,
Founder & Chairman

Neal Pronek
Chief Editor

yearBOOKS are all photo composed, color separated and designed on Scitex equipment in Neptune, N.J. with the following staff:

DIGITAL PRE-PRESS
Michael L. Secord
  Supervisor
Robert Onyrscuk
Jose Reyes

COMPUTER ART
Thomas J. Ceballos
Patti Escabi
Sandra Taylor Gale
Candida Moreira
Joanne Muzyka
P. Northrup
Francine Shulman

Advertising Sales
George Campbell
  Chief
Amy Manning
  Director

©yearBOOKS,Inc.
1 TFH Plaza
Neptune, N.J. 07753
Completely manufactured in Neptune, N.J. USA

What are Quarterlies?

Because keeping cichlids as pets is growing at a rapid pace, information on their selection, care and breeding is vitally needed in the marketplace. Books, the usual way information of this sort is transmitted, can be too slow. Sometimes by the time a book is written and published, the material contained therein is a year or two old...and no new material has been added during that time. Only a book in a magazine form can bring breaking stories and current information. A magazine is streamlined in production, so we have adopted certain magazine publishing techniques in the creation of this Quarterly. Magazines also can be much cheaper than books because they are supported by advertising. To combine these assets into a great publication, we issued this Quarterly in both magazine and book format at different prices.

Distributed in the UNITED STATES to the Pet Trade by T.F.H. Publications, Inc., One T.F.H. Plaza, Neptune City, NJ 07753; distributed in the UNITED STATES to the Bookstore and Library Trade by National Book Network, Inc. 4720 Boston Way, Lanham MD 20706; in CANADA to the Pet Trade by H & L Pet Supplies Inc., 27 Kingston Crescent, Kitchener, Ontario N2B 2T6; Rolf C. Hagen Inc., 3225 Sartelon St. Laurent-Montreal Quebec H4R 1E8; in CANADA to the Book Trade by Vanwell Publishing Ltd., 1 Northrup Crescent, St. Catharines, Ontario L2M 6P5 ; in ENGLAND by T.F.H. Publications, PO Box 15, Waterlooville PO7 6BQ; in AUSTRALIA AND THE SOUTH PACIFIC by T.F.H. (Australia), Pty. Ltd., Box 149, Brookvale 2100 N.S.W., Australia; in NEW ZEALAND by Brooklands Aquarium Ltd. 5 McGiven Drive, New Plymouth, RD1 New Zealand; in Japan by T.F.H. Publications, Japan—Jiro Tsuda, 10-12-3 Ohjidai, Sakura, Chiba 285, Japan; in SOUTH AFRICA by Lopis (Pty) Ltd., P.O. Box 39127, Booysens, 2016, Johannesburg, South Africa. Published by T.F.H. Publications, Inc.
MANUFACTURED IN THE UNITED STATES OF AMERICA
BY T.F.H. PUBLICATIONS, INC.

Classification & Ecological Habitats

The vast majority of North and Central American cichlids belong to the genus *Cichlasoma* of the tribe Cichlasomini, of the subfamily Tilapiinae, of the family Cichlidae. With no primary fishes to compete with, they readily colonized new areas, adapting to the biotopes available. There are approximately 125 *Cichlasoma* species, a number so large that subgenera were utilized in order to organize them into handleable groups. In 1908, Regan divided *Cichlasoma* into five sections which he considered as being natural groups. In 1966, Miller modified these groups, dividing them into seven sections: *Amphilophus* (formerly *Astatheros*), *Archocentrus*, *Herichthys*, *Parapetenia* (now *Nandopsis*), *Theraps*, *Thorichthys*, and *Paraneetroplus*. Further improvement is needed in the systematics of this rather large genus.

For a discussion of the confused status of the generic name *Cichlasoma*, see Burgess and Walls, *Tropical Fish Hobbyist*, 40(7):90-96 (March, 1992). *Cichlasoma* as used in this book equals the usage "*Cichlasoma* " of many recent authors.

Recently, some of the cichlid fishes of Africa have undergone comprehensive scientific revision and taxonomic divisions. The results may not be perfect, but they have led to a much clearer understanding of the genera and their species and their relationships.

The bodies of most *Cichlasoma* species are generally high and stretched out (perch-shape), laterally compressed, and covered with moderately large, usually ctenoid scales. The lateral line usually consists of two parts. They have one set of nostrils. The mouth is small to medium and the jaws are set with rows of small conical teeth, which in the outside rows are somewhat larger and usually have a typical incisor shape. There is only one dorsal fin with usually 14-19 spines, and one anal fin with usually 4-12 spines. The pectoral fins are symmetrical, with 12-18 rays, and the ventral fins are positioned underneath and a little behind the pectoral fins. The caudal fin is usually truncated or rounded.

Five other genera with one species each are represented in North and Central America as follows: *Aequidens coeruleopunctatus*, *Geophagus crassilabris*, *Herotilapia multispinosa*, *Neetroplus nematopus*, and *Petenia splendida*.

To understand these cichlids from Mexico and Central America, one must remember Nature's grand scheme and how they fit in. They are part of the ecological food chain and provide nourishment for larger predatory fishes, birds, reptiles, small mammals, and man. They in turn feed upon plants and their debris, *aufwuchs*, aquatic insect larvae, crustaceans, other small fishes, terrestrial insects, and other living organisms. Their diets change with the seasons as different proportions of their selections become more or less available. The different

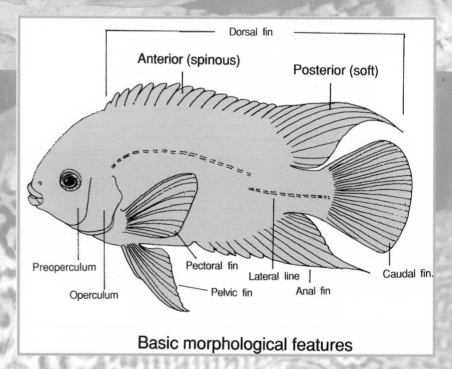

Basic morphological features

subgenera of *Cichlasoma* have particular dietary requirements which will be dealt with in those chapters dealing with the individual fishes themselves.

One must also have a brief insight into the local geological formations. The Central American and Mexican land bridge is a relatively narrow strip of land characterized by a connecting mountain range, forming the Continental Divide. Rivers rise in highlands and flow downhill into inland drainage basins that empty into the Pacific Ocean to the west of the divide and the Atlantic (Gulf of Mexico and Caribbean Sea) to the east. All twelve genera and subgenera are lowland inhabitants, with very few species found above 500 meters elevation. Above this elevation the aquatic biotope available is generally in the form of upper reaches of narrow mountainous streams and straight river channels which have quite swift currents and are too cold in temperature to support cichlid life on a permanent basis. Cichlids extensively inhabit the lowland tropical biotopes and colonize a wide range of diverse habitats. A river valley's middle section is distinctively different because of its gentler gradient. The rivers begin meandering downstream as the steepness of the mountain slope diminishes. The stream flows over sediment it has previously deposited and no longer cuts down into the rock below its bed. Instead, it erodes from side to side, gnawing away at its banks, so as to flatten and broaden the valley floor. Erosion occurs on the concave

or outer bank of the bends and sediments are deposited on the convex or inner banks. It is here, below 500 meters elevation, that we first encounter cichlids. In their lower courses, many rivers empty into lakes formed by earth movements, volcanoes, and soil erosion. A common way of classifying lakes in regard to water quality is by their individual productivity. A few impediments are the climate, the geological nature of shape and depth of the drainage basin, the chemistry of the

bottom sediments, and the nutrients brought in from the outside environment. Lakes commonly are divided into three categories: oligotrophic, mesotrophic, and eutophic, and are based on organic cycles and their nutritional or trophic characteristics. Almost all of the lakes of Mexico and Central America are eutrophic, or "well fed," and have large supplies of nutrients and heavy layers of organic sediments on their bottoms. Inflow of sediments makes most of the lakes

Meandering river formations

A Erosive surface flow
B Concave bank
C River cliff
D Shingle
E Convex bank

shallow and turbid. Eutrophic lakes are common in these regions of fertile soils, and this process is accelerated by man through urbanization and agricultural practices. These lowland bodies of water are much warmer than the rivers because of their voluminous size and ability to retain heat from the tropical sun. Paleolimnologists, those who study the geological history of lakes, generally believe that lakes evolve naturally from oligotrophic to eutrophic stages.

Plant production, the basis of the food web, is dependent on light, heat, and nutrients. A healthy ecosystem starts with the penetration of the lake's waters by sunlight, and the degree of penetration depends on the degree of transparency. Tropical lakes with high water temperatures show little seasonal temperature change at the surface or in the depths, even though they can become just as stably stratified as temperate lakes. The sun-warmed upper layers of a stratified lake are known as the epilimnion.

Lakes are not permanent features of the earth's landscape, as all will change and most eventually disappear. The process may be relatively fast because of rapid siltation, or it may be slow, occurring over hundreds of thousands of years. Disappearing lakes have produced many of the wetlands in Mesoamerica. Marshes are low-lying lands that flood when rivers overflow. Many dry up during prolonged periods of drought. Swamps are places that are always waterlogged and often brackish.

Neotropical Cichlid Habitat

Upper river valley

A

Middle river valley

B

C

Lower river valley

3

4   D

Lakes and Lagoons

Neotropical cichlid habitat showing cross-sections of upper (A), middle (B), and lower (C) courses of a typical river.

In their lower courses, some other large rivers flow down a gradient as slight as a few feet, or even inches, per mile. They bevel off the hills and empty into the sea, where deposits of sediments accumulate. Over time many river mouths become blocked with large loads of sediment and produce deltas because the tides and currents cannot carry it away fast enough. These low-lying flood plains are thickly carpeted with sand and mud, and this deposition divides the river into several channels called distributaries. These distributaries are flanked by levees and separated by lagoons. A mixture of the more salt-tolerant species is found here.

Under riverine conditions cichlids comprise about 25% of their biotope compared with the cypriniforms, which make up approximately 40%. Under lacustrine conditions they comprise about 30% of the biotope or about the same percentage as the cypriniforms.

Reproductive Behavior in Nature

Mating systems and the degree of care provided by the individual parents is determined by the numerous environmental factors that influence male and female reproductive success. Some species experience a range of environmental conditions, while others have a much narrower margin of adaptability. However, mating and parental strategies of most species are sufficiently flexible, showing adaptive modification in relation to environmental changes.

These oviparous fishes are basically monogamous substratum spawners or substratum brooders with a biparental care system for their eggs and fry. All these cichlids practice intense care over their young until they are mature enough to safely search for food on their own. However, inspection and observation of cichlid reproductive behavior in their natural habitat is not an easy task. They are secretive during this time and prefer to breed in quiet, undisturbed locations, usually around submerged trees and the hollowed out caves beneath the river's concave banks.

The normal breeding season is during the wet months, when the temperature is warm, the water clean, the food abundant, and adequate safe and shallow breeding sites are available. Males tend to select their mates from the colony present, find a suitable substrate on which to lay their eggs, and begin bonding and territory formation. Females naturally grow to smaller sizes than the males and prefer to mate with the largest males of the population. This is partially because the largest or most robust males tend to show their dominance and superiority through intensified, brilliant coloration. The males of some *Cichlasoma* species even produce a prominent nuchal hump. The females exhibit a maternal reproductive pattern, many with black pigments throughout the body in patterns of splotches or lines. Thus, courtship proceeds with a breeding dress, usually by both species, and is followed by aggressive displays from the male, who often slaps bodies and tails with its selected mate while circling in a tight area with great speed. The male's body quivers violently, somewhat like a physical convulsion, with fins erect. After a couple of hours his genital papilla drops farther. Substratum type preferences vary from one group to another in their natural environment. Whether rock, wood, or sand, the spawning medium, or spawning site, is laboriously cleaned by both sexes. The female normally deposits her eggs within 3-5 days after the onset of courtship. The spawn sizes vary from one species to another, but the eggs are laid in irregular, circular or oval formations or plates, conforming to the size and shape of the chosen spawning medium. The female will lay a row or two of eggs at a time, trading places with the male who directly follows in tandem, releasing his sperm to fertilize the eggs before their casings harden. The female passes over the spawning site again and again, laying more and more eggs, only to be followed again by the industrious male. This process usually is completed in 2-4 hours, depending on the amount of eggs laid and whether or not any outside competition or interference exists.

Generally, the female does most of the hands-on work, rarely straying more than a meter from her brood, controlling the immediate center of their territory. She stays quite busy from the time she lays her eggs until she relinquishes control of the juveniles. On the other hand the male stands guard and protects the outer perimeter of their territory.

The eggs constantly are scrutinized so as to keep away any loose debris. A forward push or fan of oxygenated water by the pectoral fins removes most of the bacteria which might adhere to the egg and deteriorate its sheath. A few unfertilized eggs will turn white and fungus within a few hours. They are removed by one of the parents to prevent further contamination. The female performs the major part of this action, while the male positions himself nearby, controlling the outer limits of their temporarily expanded breeding territory. The smaller female occasionally will trade places with the larger male and patrol this combat zone while the male assumes her responsibilities. This area can be as small as a cubic meter for smaller species or as large as 4-5 cubic meters for the larger varieties.

The fecundity or total number of eggs produced varies from species to species depending upon the maturity of the female. The older the female, the greater the number of eggs produced, assuming she is well conditioned. Under the artificial conditions of captivity, some females continue to produce for 7-8 years. However, it must be remembered that the longevity of captive specimens can be considerably longer than that of those existing in the wild. Survival and competition is indeed a daily task in their natural environment. As aquarists, we can help to ensure the survival of the many fragile species that frequently are victimized in their native third world environment.

The majority of the riverine species adhere their eggs to the submersed rocks that lie beneath the overhanging banks. Some prefer to lay their eggs on exposed roots and submerged trees that line the river's edge. Here, the inner banks, or concave banks, are worn away by the current and honed out underneath so as to form caves in which the fishes can find shelter from other aquatic predators and the blistering sun that parches

this area of the world. Brooding parents easily can mark off this territory and defend it during the period in which their fry are immobile and defenseless. This concealed area is also a safe haven from predatory birds and animals.

During the springtime, some of the small, shallow drainage streams and creeks become day-care centers and temporary homes to many brooding females and their spawns. The female always keeps an immediate space or territory secure for her fry and will defend it ferociously. She interacts on a compatible basis with the

Cichlasoma (Amphilophus) rostratum. Lago de Nicaragua, Nicaragua.

other brooding females and their juxtaposing territories, while leading her young up and down the banks in constant search of food. The slow-moving current allows the female to more easily tend her brood during their early stages of growth. The phytoplankton and zooplankton necessary for fry development are also more abundant in this warm, shallow biotope. Some riverine *Amphilophus* and *Theraps* species even practice communal brood care called **creching**, whereby the different fry from the same species in the immediate area will congregate together

under the watchful eyes of the tending females. Some juveniles even will trade parents and let them do temporary babysitting. I have witnessed this parental care system in colonies of C. (Amphilophus) rostratum, C. (A.) altifrons, and C. (Theraps) nicaraguense in their native habitat. This never could happen with the more conspecific, lacustrine species. The fry of some species feed upon the parents' protective slime cover until large enough to feed on the other microscopic food sources available. After the fry reach the juvenile stage (about 10-20 mm and 2-4 weeks) they rapidly gain size and mobility. The parents lead them across the river to feed in the shallows of the outer or convex banks. This area represents the sand bank of the river and is called a **shingle**. It consists of gravel, pebbles, and small particles of sand. The males abandon the family after a few weeks and leave the rearing to the females. Day by day the juveniles expand their teritories, and the tight shoal of small fry becomes a struggling band of survivors reduced in numbers due to predation and disease. By the time the youngsters disband and venture out on their own, their numbers are diminished greatly to about 20-30 percent of their original count.

The smaller *Archocentrus, Thorichthys,* and *Herichthys* species patrol a smaller area of about 1 cubic meter or less. Many of the smaller species prefer the smaller tributaries to breed and raise their progeny. These slower moving waters are found in the lower river valleys. These smaller sized *Cichlasoma* species breed often and at young ages, indicative of the high degree of predation upon these species.

If lacustrine in nature, the majority of the cichlid breeding territories are found against the shoreline among the larger strewn boulders and submerged trees. I have observed the majority of their breeding sites or nests to be between 1 meter and 3 meters deep. The distance between the breeding pairs differed according to subgeneric types. The *Nandopsis* or guapote control large areas, between 3-4 cubic meters. The lower river and lacustrine *Theraps* and *Ampilophus* varieties prefer to dig sand pits or "arenas" in water approximately 1-2 meters deep. Their pits are about the size of a basketball and are generally dug against some hard substrate, such as a rock or a piece of wood. The *Theraps* and *Amphilophus* species, in constant vigilance and protection of their young, patrol a medium sized territory of about 2 cubic meters.

Among the biparental cichlids, there are reports of natural polygamous populations of *C. (Archocentrus) nigrofasciatum* and *Aequidens coeruleopunctatus*. I have observed this phenomenon in *C. (Archocentrus) sajica* and *C. (Archocentrus) septemfasciatum* (the southernmost variety from the Rio Sixaola drainage). What environmental factors are responsible for male desertion are not certain.

Aquarium Management

All of the cichlids of the world could theoretically be bred in captivity, and the North and Central American complex offers the aquarist a wide range of breeding electives and possibilities. Both the novice and the experienced aquarist can find their efforts rewarding, with some introspection as to how these fishes should be maintained properly. Whatever one's taste might be in cichlids, it is important to realize that these colorful and interesting freshwater fishes have individual requirements. Most can be kept easily if they are maintained intelligently. Community tanks can be established for all species if they are kept in sensible groupings according to densities, sizes, behavioral traits, and similar water and dietary requirements. By doing your homework before choosing the pets of your choice, a considerable amount of confusion and frustration can be avoided.

The most compatible community aquariums are those that are filled with juvenile residents of the same age and raised as a family unit. A normal pecking order will develop, but it will be of a less aggressive nature than if a few large adults were introduced together for the first time. As they grow up and interact together on a day-to-day basis, they form a small family structure. This allows the fishes lower down in the pecking order a better chance of survival, assuming an adequately sized aquarium with a sanctuary of hiding places is provided. An aquarium of between 50-75 gallons minimum conveniently can house most community residents. The larger the aquarium, the better off the fishes will be, as maintenance can be reduced with no ill effects on the residents. Remembering the conspecificity of these individual species, a good mix would be of similar sized species from the different genera and their subgroups.

The same concept of procuring small fishes for the purpose of raising and breeding should be followed. A minimum of six juveniles should be purchased if the ultimate goal is to have even one breeding pair in a year's time. The Central American complex basically can be divided into three groups of small, medium, and large sized species. This is the basic premise with which to group and house these species. A standard rule of thumb for breeding these fishes is about 10 gallons for each inch of fish measured at an adult size. The small species should be housed in 20-30 gallon aquaria. The subgenera *Archocentrus* and *Thorichthys* and *Herotilapia multispinosa* are tailor-made for this size aquarium. These types grow to about 2.5-4 inches as adults and are the dwarf cichlids from this section of the world. As bottom foragers, they browse over the bottom substrate and sift through the detritus. These type feeders are the most docile, the easiest to maintain, and the easiest to breed.

They are good selections for the beginning cichlid enthusiast. They are also good for all aquarists who already have small aquaria that require smaller fishes. Many people have a budget to live on and grandiose aquaria are not feasible or possible. The smaller tanks and equipment are cheaper and indoor space can be a premium with today's apartment living. The enjoyment of nurturing an adult pair, having them spawn, and watching them rear their progeny through their several stages of growth can be just as rewarding and educational as when breeding the larger, more difficult varieties.

The majority of North and Central American varieties fall into the medium category. Included are the smaller members of the subgenera *Amphilophus* and *Theraps*, as well as *Neetroplus nematopus*. This group of fishes grow to an adult size of about 4-7 inches and are generally of average temperament as far as cichlids go. The medium sized species should be housed in aquaria of 40-75 gallons. Some are easy to breed and some are rather difficult to breed under captive conditions. There now exists an ample and ready market in which to sell or trade your young, the fruits of your

efforts.

The larger species of *Amphilophus, Nandopsis*, and *Theraps* should be maintained in aquaria of 75-125 gallons capacity. They can be raised properly in small aquaria, increasing in size as they grow larger and older, but ultimately a large aquarium will be required to breed them. Even these large aquaria offer dimensions of far less size than their territorial dimensions under natural conditions. During their fierce courtship, any other cichlids

Cichlasoma (Amphilophus) alfari.
Rio Canas, Costa Rica.

should be removed for self-preservation, as a spawning pair will disrupt most all aquaria. They desire to be free of any potential predators and the pair eventually will kill off the less dominant, one by one. This belligerent behavior is the key to their survival in their natural habitat, and the aquarist should deal with the situation in a prudent and humane manner.

The water of most all of Mexico and Central America is alkaline and hard. The pH averages 7.0-8.0, and the hardness ranges are generally between 3-25 degrees DH. Most of us in the U.S. and Europe have similar water

conditions, and these fishes can tolerate one degree of pH change and 5-10 degrees of hardness, DH, change with little adversity whatsoever. It's easier to raise the pH with dolomite or crushed coral than it is to lower it with peat bags or similar chemical treatments. These Mesoamerican cichlids generally accept a slightly higher pH. Do keep in mind that ammonia is more harmful in alkaline water than in acidic water.

Many lowland rivers and spring-fed creeks have a neutral pH and can be a bit softer depending over which substrate they are flowing. The rivers in the mountainous highlands generally have a higher pH, DH, oxygen, and mineral content than those that run through the lower flatlands with more fertile soils, or the lakes. The physical and chemical water characteristics of the individual species' habitats are discussed in the chapters on subgenera and species. Whatever your water chemistry is, set up and fill the tank at least a week before you put in the new residents.

Second in importance to the size of the aquaria is an effective filtration system, a must to remove unwanted waste products and purify the water. Several mechanical filters exist that adequately can perform this function. The size and capacity requirements should match at least the minimum specified recommendations for the size of aquarium used. However, one must take into consideration the degree of difficulty of the routine maintenance required when purchasing such a system. A filter with easily replaceable packs or cartridges can make maintenance much easier and less time consuming.

Know the natural environment and regions from which your cichlids come and try to recreate them as closely as possible (a biotope aquarium). Riverine fishes prefer fast flowing waters and require a high capacity filter and pump to thrive and breed under artificial conditions. Their water should be changed a bit more often than when dealing with the lacustrine species which can tolerate higher concentrations of ammonia, nitrates, and nitrites.

Undergravel filtration is very important, as the bacteria that exist in the substrate perform the natural process of nitrification. A minimum of 3 inches of pea-sized gravel (3/16 of an inch diameter) should be placed in the aquarium when managing small and medium sized cichlids and a minimum of 4 inches of a coarser aggregate should be used when dealing with larger cichlids. They can do some serious digging and moving of gravel so a thicker level is needed to keep the system functioning properly. Slope the gravel so it is elevated toward the back of the aquarium for easier maintenance and management. Electrical power heads should be used on the uplift tubes to ensure better filtration and create the effect of flowing or moving water. The additional investment is well worth it when it comes to managing and breeding these magnificent fishes.

The overhead lighting can be fluorescent or tungsten and should be set for about 10-12 hours regular use and increased to about 14 hours per day during breeding times. The heater (approximately 2 watts per 10 gal. water) should be set for about 76-79° F most of the time and raised to 80-83° F during breeding times.

Rockwork of different sizes and shapes is okay for decorative as well as functional use. Generally, refrain from the use of caves for all but the smaller Latin American species, as they tend to encourage aggression. You still can make the aquarium private for them by placing plants (real or artificial) or driftwood toward the front and middle. When using live plants, stick with the *Anubias, Echinodorus*, and *Microsorium* species, and any other thick-leafed varieties, or else they will be devoured over a short time. Some aquatic plant enthusiasts place about a one-inch-square wire mesh in the bottom of the aquarium on about 1 inch of gravel. They install the plants through the square holes and then fill the aquarium up with the necessary gravel amount (3-4 inches). This prevents the plants from becoming pulled out by all but the largest of the cichlids, and as they continue to grow and root out they become firmly entrenched by the wire. Encourage the pair you plan to breed to lay their eggs on a smooth flat rock of darker color, by placing it slightly off center and toward the back of the aquarium. Tilt the flat rock at a slight grade, 15-30 degrees, toward the nearest back corner. This does not ensure they will pick that rock as their spawning site, but they usually do if the rock is approximately 1 1/2-2 times the size of the adult fish.

Avoid the tendency to overfeed these fishes, and alternate the food types with varied food sources. Use a large-sized fish flake or fish pellet as a supplemental feed. Most all the larger, reputable dealers sell a cichlid line of both food forms. Additional supplements of green vegetable matter should be added to their diet on a regular basis. This can be done with a spirulina or algae-based flake or pellet. Frozen spinach can be shaved with a knife and is readily accepted after a few trial feedings. If they don't eat it the first few times it is offered, siphon it out of the aquarium and try holding them off of any food sources for a couple of days. This usually will induce them to eat it willingly. If not, try blanching other green vegetables, such as zucchini or lettuce, as it is most important to keep these natural minerals and vitamins in their diet. Also, live foods should be administered for reproductive conditioning. Several types of worms (live are preferable) are available readily and have provided good results, even for fishes that are most reluctant to breed.

Target fishes also should be placed in the aquarium to modify the pair's aggressive behavior with a bonding approach. These fishes should be small, common cichlid varieties, and it should be realized their lives probably will be short because of the spatial and territorial demands of a brood pair. The savagery or ferocity of a bonded pair during the brood season almost is unmatched in both the freshwater and marine fish world. It is a normal behavior for self-preservation in the cichlid's natural habitat and should be recognized and appreciated by the aquarist.

Other methods for managing this aggressiveness, which is usually by the male during the prespawning or courtship period and by the female during the postspawning period, can be that of the incomplete divider, whereby a plastic grating is placed between the two fish and centered on top of a large, flat rock. Assuming the female lays her eggs on or near the centered rock, a portion of her eggs will be fertilized without risk of any injury to either partner. The separate compartment method also is useful if one member of the pair is significantly larger than the other. Holes cut out of plastic grating should be just large enough to allow only the smaller member to swim through. Generally, the smaller member is the weaker member and can attain sanctuary within the confines of the same tank, provided this method is used.

At hatching time, the parents chew the wrigglers out of their eggshells and continue to tend them until they are mobile. They are tended to by both parents, but I have found it best to allow only

the female to handle this chore. She will have no problem in this safe, artificial ecosystem, and the chance of an overnight quarrel between the pair is eliminated. The fry of several species eat the mucus or slime from the parents' sides until old enough to accept other foods. Live *Artemia* or brine shrimp is the easiest and most reliable food to raise the small fry upon, producing excellent growth rate results. Again, be careful not to overfeed, as they can contaminate an undergravel bed if left in too large a concentration.

Cannibalism can occur from time to time but is usually followed by a fairly short convalescent period and another breeding by the two adults. Young breeding pairs are most apt to exercise this phenomenon, but usually will stop by the second or third spawning. If not, remove the parents after the eggs have been laid, and place an airstone near the slightly

elevated rock. This will keep the bacterial debris from coming into contact with the eggs by the motion of the air bubbles. Also, administer methylene blue to prevent microbial infestation.

Fry grow much faster if not overcrowded, so if a spawn is large, divide it up into two tanks about the third or fourth week. You will see immediate and amazing results. Be sure to fill the new, second tank with about 50% of their original tank water and the other 50% with new water that has been aged for two or three days. In 3 to 4 months the fry should be about $1\frac{1}{2}$-2 inches in total length and ready for sale or trade. If you manage to raise successfully only about 2/3 of the original spawn, don't worry. You've done a very good job, and this represents as good a survival rate as can be expected. These numbers are far greater than the percentages of survival in the wild.

Cichlasoma (Amphilophus) longimanus. Rio San Juan, Nicaragua.

Amphilophus

Type species: *Cichlasoma labiatum* (Guenther, 1864)

This large group of cichlids originally was referred to as *Astatheros* Pellegrin, 1904. In 1966, Dr. Robert Miller restructured this group and named it *Amphilophus*. This subgenus has medium to large sized fishes with rather deep bodies, straight snouts and slanted or sloped cranial profiles. Their mouths are of small to average size and their jaws have standard protractability. Their outer series of teeth are numerous in both jaws. Their dorsal fins have 10-15 rays and the anal fin has 8-10 rays.

This grouping of cichlids ingests food from open areas by sifting through the gravel or sand substrate. They tilt their body and plunge their snout into the substrate, returning to their normal horizontal position to agitate the contents. They sift and strain the bottom detritus for edible objects and spit out the remaining components. They also consume stands of algae and nibble or peck away at the *aufwuchs* that grows upon the hard substrate. They are opportunistic feeders, devouring insects, snails, and other small fishes that are left unprotected by their parents. As with most Neotropical cichlids, breeding is more frequent during the wet season than during the dry. Adult pairs set up territories in the shallows around banks or the shoreline, among the larger boulders and submersed trees. They proceed to breed and raise their fry in the safety of the many small caves and crevices formed by these rocks and trees. Their territories encompass about 1-2 cubic meters, from which all predators are repelled. Courtship behavior is aggressive, as the male encourages a selected female to a suitable breeding site with a dancing ritual. His colors brighten considerably as he displays his extended fins and his body shakes and quivers while he circles around or next to the selected site. Head bumps, body nudging, and fin nipping are all part of this courtship behavior. A few days later, about 700-2000 eggs are laid on the oblique or vertical walls of the cave, where they constantly are attended to by both sexes. About three days later, the wrigglers are chewed from their egg casings and kept together in a compact group or pile. The fry feed upon infusoria and phytoplankton and supplement their feeding by eating from the dermal mucus on their parents' bodies for several days. The males soon move on, leaving the females alone to tend the school for several more weeks. Communication between the parent or parents and the fry is by a series of signals whereby head, body, and fin jerks are employed. In response, the fry gather tightly together and school close by the parents. At approximately 1-2 months of age the juveniles become more mobile and increasingly move out into the open. The numbers of young diminish greatly as this family disbandment takes place due to predation. Generally, less than 20% live to maturity.

AMPHILOPHUS: GROUP 1

This subgenus has two major phenotypes, distinguished by their feeding patterns. Members of the first group are larger and more robust and are generalized predators. They have short snouts and average to wide mouths. Their caudal fin is rounded or subtruncate. They are substrate spawners and prefer caves and crevices in the rocks for shelter.

C. (A.) citrinellum (GUENTHER, 1864)

SYNONYMS: *Heros citrinellus* Guenther, 1864, *Heros basilaris* Gill and Bransford, 1877, *Cichlasoma granadense* Meek, 1907

COMMON NAME: Midas cichlid; its native Spanish name is *mojarra rayada*.

DISTRIBUTION: Nicaragua and Costa Rica, on the Atlantic slope. The Great Lakes and the crater lakes of Nicaragua to the Rio Matina in the central coastal plains of Costa Rica.

HABITAT: pH: 7.0-8.75, GH: 3-18, KH: 3-21, T: 23-33° C. Lacustrine in nature, this

Cichlasoma (Amphilophus) citrinellum. Lago de Nicaragua, Nicaragua.

species is distributed widely in the lakes of Nicaragua and is the most common cichlid in the lakes where it is found. They are uncommon in the rivers but will penetrate the lower river valleys where the water is slow flowing or tranquil. They rarely venture into the faster moving waters of the highlands. These fishes are omnivorous, eating mostly *aufwuchs*, insects, snails, and small fishes.

DISTINCTIVE CHARACTERISTICS: SL (standard length): 244 mm. The majority of these fish are of the normal cryptic coloration (black, gray, or brown), matching the substrate for camouflage and survival purposes. They communicate with each other by changing their markings, either adding or removing bars or spots. However, about 10% of this species is xanthomorphic, undergoing a color metamorphosis at varying stages of growth. Gradually, bright colors emerge, ranging from white to yellow to orange. Black blotches or patches occur on many specimens,

Cichlasoma (Amphilophus) guija. Illustration by John R. Quinn.

generally on the fins and lips, but these usually disappear with age. During pair formation the males become sexually dimorphic as their foreheads swell, forming a pronounced nuchal hump.

C. (A.) guija HILDEBRAND, 1934

DISTRIBUTION: Southern Guatemala and El Salvador on the Pacific slope. From the Rio Lempa to the Rio Paz in SE Guatemala.

COMMENTS: Some illustrations portray this fish as having fleshy lips, much like those of *C. altifrons*.

C. (A.) labiatum (GUENTHER, 1864)

SYNONYMS: *Amphilophus froebelii* Agassiz, 1859, *Heros*

Cichlasoma (Amphilophus) labiatum. Lago de Nicaragua, Nicaragua.

labiatus Guenther, 1864, *Heros lobochilus* Guenther, 1869, *Heros erythraeus* Guenther, 1869, *Cichlasoma dorsatum* (Meek, 1907)

COMMON NAME: Red devil; its native Spanish name is *mojarra picuda*.

DISTRIBUTION: Nicaragua, on the Atlantic slope. In the Great Lakes of Nicaragua: Lago Managua and Lago Nicaragua.

HABITAT: pH: 8.5-8.7, GH: 3-4, KH: 3-4, T: 28-33° C. Same as *C. citrinellum*. This species is found sympatric with *C. citrinellum*, but rarely penetrates into the rivers, accounting for its limited geographical distribution.

DISTINCTIVE CHARACTERISTICS : SL: 240 mm. Very similar to that of its sibling species, *C. citrinellum*, except that its lips are thicker and larger, its head is thinner and more pointed, and its fins are shorter. Its body is also more slender and elongated.

As in *citrinellum*, about 10% are polychromatic but of a brighter hue of pink or red and often with irregular black patches.

Cichlasoma (Amphilophus) lyonsi. Photo by Hans J. Mayland.

C. (A.) lyonsi GOSSE, 1966

DISTRIBUTION: Costa Rica and Panama, on the Pacific slope. From the Rio Cota in southern Costa Rica to the Rio Dupi in Panama.

HABITAT: pH: 8.2-8.6, GH: 5-6, KH: 5-6, T: 26-29° C. It inhabits rivers with moderate flow and sandy bottoms. This fish feeds basically upon seeds, aquatic insects, and bottom detritus.

DISTINCTIVE CHARACTERISTICS : SL: 150 mm. The body is of a dark gray or dark brown color with seven square black patches on its side between the operculum and its tail and an extra one at the base of the caudal. The spot in the caudal stands out because it is encircled by a bright yellow ring. The squares are overlaid

on seven bars which are hardly visible on adults, but well defined in the juvenile pattern. The body sides and fins are a yellowish orange in adults.

The dorsal and anal fin spine counts are low, 15 and 6 respectively.

C. (A.) zaliosum BARLOW, 1976

COMMON NAME: Arrow cichlid

DISTRIBUTION: Atlantic slope of Nicaragua. Endemic to Lake Apoyo.

Cichlasoma (Amphilophus) zaliosum. Illustration by John R. Quinn.

HABITAT: pH: 8.1

DISTINCTIVE CHARACTERISTICS : SL: 162 mm. The body coloration is a greenish gray to blue with black markings. When breeding, there are seven black vertical bars on a pale body, each bar shaped slightly like an hourglass.

AMPHILOPHUS: GROUP 2

Members of the second group are medium sized substratum-sifting invertebrate feeders. Their bodies are not as deep and are more slender than the previous group. Their snout is rather long for scooping into the sand in their never-ending quest for food. The caudal is subtruncate or emarginate with rounded lobes. They spawn in depressions in the sand or gravel and are basically more delicate in nature than their more robust cousins. Their spawns are about one half the size, with between 300-1000 eggs per spawn.

C. (A.) alfari MEEK, 1907

SYNONYMS: *C. alfaroi* Meek, 1907, *C. lethrinus* Regan, 1908, *C. bouchellei* Fowler, 1923

COMMON NAME: Pastel cichlid

DISTRIBUTION: Honduras to Panama, on both Atlantic and Pacific slopes. Ranges from the Rio Patuca in Honduras to the Rio Guarumo in Panama on the Atlantic slope. A population also exists on the Pacific slope in the northwestern province of Guanacaste in Costa Rica.

HABITAT: pH: 6.6-7.6, GH: 3-8, KH: 4-8, T: 20-34° C.

DISTINCTIVE CHARACTERISTICS: SL: 150 mm. This riverine species inhabits most all sections of the river, from the fast flowing upper sections to the slower moving lower reaches. They are found up to 1150 meters of elevation and prefer habitats that have an excess of aquatic insects to supplement their diet. The cranial profile has a sharp

Cichlasoma (Amphilophus) alfari. Female. Rio Cuba, Costa Rica.

Cichlasoma (Amphilophus) alfari. Rio Sixaola, Costa Rica/Panama.

slope with a rather long snout, and the upper jaw extends slightly over the lower jaw. There are three types, all of which have their dorsal fins tipped or edged in red. Their pelvic fins, anal fins, and the bottom half of their caudal fins are free of any spangling. Females have a black elongated spot in the middle of their dorsal fin. About 300-400 eggs are laid at the end of the dry season, just before the wet season.

TYPE 1: Atlantic variant collected in the Rio Cuba near Limon, Costa Rica. A dark line runs between the eye and the caudal spot. Six vertical bars are present, but faded below the lateral line. This population has a jade green body coloration, with deep pink breast and underside. Irregular turquoise spangles cover the lower face and extend just above and behind the gill cover. This same coloring borders the scales throughout most of the body, giving it a reticulated pattern. The face and pelvic fins are of a bright butterscotch coloration.

TYPE 2: Atlantic slope variant found in the Rio Sixaola drainage in southern Costa Rica and northern Panama. This population is more elongated and not as deep bodied. An interrupted lateral line extends the length of the body. Six wide vertical bars are present, with the first and second bars uniting and forming a "Y" pattern. It has an olive-yellow body coloration with 14-15 horizontal rows of small black spots or dots covering the entire body. Three or four blue streaks are seen below the eye, and no spangling is found on the face,

Cichlasoma (Amphilophus) alfari. Rio Sixaola, Costa Rica/Panama.

head, or gill cover. Very little spangling or spotting is found on the dorsal fin, and none of the other fins are spangled whatsoever. This may very well be a new, undescribed species.

TYPE 3: This Pacific slope variant, from northwestern Costa Rica, has a yellow body coloration with pink undersides and lips. They have much less body spangling, but much more fin spangling. The typical blue-colored scale bordering is noticeable through the middle and posterior sections of the body. The operculum is spangled, but very little spangling is found on the face.

C. (A.) altifrons (KNER & STEINDACHNER, 1863)

DISTRIBUTION: Costa Rica and Panama, on the Pacific slope. From the Rio Terraba in southern Costa Rica to the Rio Chiriqui in Panama.

HABITAT: pH 7.7-8.2, GH: 2-5, KH: 2-6, T: 22-29° C.

DISTINTIVE CHARACTERISTICS: SL: 130 mm. This species has 5 wide black bars on a bright yellow body with aquamarine spangling over its body and fins. Its upper jaw protrudes over the bottom jaw, and the lips are thickened, indicative

Cichlasoma (Amphilophus) diquis. Rio Esquinas, Costa Rica.

of their adaptability for rocky or sandy stream habitats. They are usually found in moderate to fast flowing water and between 20-400 meters elevation. They feed upon the *aufwuchs* and invertebrates found there as well as various aquatic insects. The males of this species are aggressive and intolerant of other males of their own species (conspecific). However, they get along fine with the other cichlids with whom they share the habitat: *C. diquis*, *C. sieboldii*, and *C. sajica*. It has been reported that they cover their eggs with sand, but I have never witnessed this peculiar behavior.

C. (A.) calobrense MEEK & HILDEBRAND, 1913

DISTRIBUTION: Panama, on the Pacific slope. In the Rio Bayano and the Rio Tuira.

HABITAT: pH: 7.2-8.1, GH: 3-5, KH: 3-5, T: 27-32° C.

DISTINCTIVE CHARACTERISTICS : SL: 260 mm. This species is robust and shaped similarly to *C. citrinellum*. The body coloration is bright green, similar to that of *umbriferum*, each scale being iridescent.

C. (A.) diquis BUSSING, 1974

DISTRIBUTION: Costa Rica and Panama, on the Pacific slope. From the Rio Terraba in southern Costa Rica to the Rio Parrita in Panama.

HABITAT: pH: 7.7-8.2, GH: 2-5, KH: 2-6, T: 23-33° C.

DISTINCTIVE CHARACTERISTICS : SL: 135 mm. This is the sibling species to *C. alfari*, but is not as brightly colored. Its body coloration is a straw yellow, with an olive green face and a light pink breast. Six prominent, irregular bars give this fish some contrast,

as no spangling is found on the body or face and only a little is found on the caudal or soft rayed sections of the dorsal fin. This species is more aggressive than its counterpart.

C. (A.) longimanus (GUENTHER, 1869)

SYNONYMS: *Cichlasoma popenoei* Carr & Giovannoli, 1950.

COMMON NAME: Red breast cichlid. Its native Spanish name is *viejitos*.

DISTRIBUTION: Southern Guatemala to Costa Rica, on

Cichlasoma (Amphilophus) calobrense.

the Pacific slope. From Rio Nahualte in Guatemala to the Rio Bebedero in Costa Rica.

HABITAT: pH: 6.5-8.7, GH: 3-

Cichlasoma (Amphilophus) altifrons. Rio Esquinas, Costa Rica.

Cichlasoma (Amphilophus) calobrense.
Photo by H. J. Mayland.

4, KH: 3-4, T: 23-36° C.

DISTINCTIVE CHARACTERISTICS :
SL: 135 mm. Juveniles
have a silver-green body
coloration with 8-9
vertical bars and irregu-
larly shaped spots
where they cross the
mid-body line. One bar
also crosses the fore-
head just over the eye.
Adults display a lateral
bar from the gill cover to
a prominent dark spot
which is present near
the middle of the body.
Its lips are quite thin,
indicative of its prefer-
ence for sandy bottoms
from which to feed. The
body coloration is basically
lime green with a bright or hot

*Cichlasoma (Amphilophus)
longimanus.* **Rio Canas, Costa
Rica.**

pink breast and abdomen.
Eight to nine horizontal rows
of green spangles extend from
behind the operculum to the

caudal peduncle. Its
dorsal fin has rows
of these iridescent
green spots, only
larger. The area
between the rays of
the anal fin is
streaked, and the
caudal fin is lightly
streaked on the
lower half, a reverse
to that of *C. alfari.*

C. (A.) macracanthum* (GUENTHER, 1864)

DISTRIBUTION: Southern
Mexico to El Salvador, on the

Cichlasoma (Amphilophus) macracanthum. **Rio
Motagua, Guatemala.**

Pacific slope. From the Rio
Tehuantepec in Mexico to the
Rio Paz in El Salvador.

HABITAT: pH: 6.4-7.0, GH: 4-
5, KH: 4-6, T: 26-30° C.

DISTINCTIVE CHARACTERISTICS :
SL: 250 mm. This species'
most identifiable feature is
its five wide, black bars. Its
body is rather tall and its
coloration varies slightly
from light green to white. It
is rather drab when young,
but the black and white
contrast of an adult re-
sembles that of
Cyphotilapia frontosa from
Africa.

*This species might more correctly be
placed in Group 1 *Amphilophus.*

C. (A.) margaritiferum (GUENTHER, 1862)

DISTRIBUTION: Guatemala, on
the Atlantic slope. Previously
known from only one specimen, 6 1/2 inches long, found
by Mr. Salvin at Lake Peten.

HABITAT: pH: 7.5-8.0, GH:
15, KH: 6, T: 28-30° C.

DISTINCTIVE CHARACTERISTICS :
Size 235 mm. This species is
known from a single specimen
whose exact type locality is
unknown. Guenther (1862)
recorded the original specimen
of *margaritiferum* as coming
from Lago Peten but the
catalogue of the British
Museum (Natural
History) shows nothing
under the type locality.
Many ichthyologists felt
that the species had
become extinct since
the holotype was
collected.

It is not extinct! It is
indeed alive and well
(although rare) in Lake
Peten. We were able to
capture only six of
them in an area of the
lake where a small
creek (with no name)
emptied. The temperature of
the water around the mouth of
the stream was halfway
between the warm lacustrine
and the cooler riverine temperatures, about 28° C. Bussing & Martin (1975) stated

*Cichlasoma (Amphilophus)
margaritiferum.* **Lago Peten,
Guatemala.**

that *C. (A.) margaritiferum* is an intermediate form between *Thorichthys* and *Amphilophus*. Its five sensory pores on the mandible and subopercular blotch are *Thorichthys* traits that are unique in *Amphilophus*. In fact, several other diagnostic traits are shared by the two groups.

Adult male *C. (A.) margaritiferum* display a bronze body color, each scale having a medium to large pearly white spot. The dorsal, caudal, and anal fins are spangled with white. Adult females are bright yellow, but lack the intense spotting.

indicating its mundane environment. Its body is wine- or rose-colored, with 12-13 horizontal rows of turquoise colored spots. These spots or dots continue into the rays of the caudal fin and the dorsal fin, but are slightly larger toward the anterior section of the body. An irregular, iridescent green blotch appears about the middle of the side. The females are similar, but with much less spotting.

Cichlasoma (Amphilophus) robertsoni. **Rio Usumacinta, Mexico.**

C. (A.) robertsoni REGAN, 1905

SYNONYMS: *Cichlasoma acutum* Miller, 1907

COMMON NAME: Emerald cichlid; native Mayan name is *tepemechine.*

DISTRIBUTION: Mexico to Honduras, on the Atlantic slope. From the Rio Papaloapan basin, Lago Tamascal, Mexico, to the Rio Cangrejal.

HABITAT: pH: 7.4-8.0, GH: 13-15, KH: 5-6, T: 26-30° C. Widespread throughout the Rio Usumacinta basin. Also found in the Lago Peten basin. This rather robust species prefers lower and middle sections of rivers in slower moving waters. It prefers a soft substrate of sand, mud, and small stones.

DISTINCTIVE CHARACTERISTICS : SL: 190 mm. This species is

both riverine and lacustrine, accounting for its fairly wide distribution. Those that are found in the faster flowing streams and rivers have more elongated bodies with sharper sloping cranial profiles. The lacustrine and lower river varieties have adapted with slightly larger bodies which are deeper. Their heads are not as steeply sloped or angled, and the snout does not protrude as much as those from the riverine environments. Adults are extremely colorful in their turquoise body dress, with an underlying red background in their dorsal, caudal, and anal fins. The northern races exhibit a more blue coloration, while the southern races are green. The northern races have neat rows of spots on the rays of the dorsal, caudal, and anal fins. The southern races have irregular spotting of different sizes randomly scattered throughout the dorsal and have more streaking on the gill cover and lower facial area beneath the eye. Adult pairs generally spawn in open pits or depressions in the sand. They clean an area about 12-18 inches in diameter, and patrol an area about 1-2 cubic meters, diligently

Cichlasoma (Amphilophus) rhytisma. **Rio Telire, Costa Rica.**

C. (A.) rhytisma LOPEZ, 1983

DISTRIBUTION: Costa Rica and Panama, on the Atlantic slope. In the Rio Sixaola and the Rio Telire in Costa Rica to the northern range of Panama.

HABITAT: pH: 7.4-7.8, GH: 4-8, KH: 4-6, T: 24-26 °C. This elongated species prefers moderate to fast flowing rivers and feeds on aquatic insects and bottom detritus.

DISTINCTIVE CHARACTERISTICS : SL: 140 mm. *C. rhytisma* is relatively rare in nature. Its head is triangular in shape,

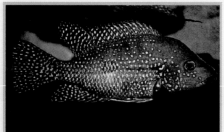

Cichlasoma (Amphilophus) rostratum. Lago de Nicaragua, Nicaragua.

warding off all intruders. The female's breeding dress is not as brilliant as when under normal conditions. Instead, she turns a drab green and a prominent bar pattern appears, warning all others to keep their distance. This species sifts through the substrate, ingesting all edible foods and spitting out everything else.

C. (A.) rostratum (GILL & BRANSFORD, 1877)

NATIVE SPANISH NAME: *masamiches*

DISTRIBUTION: Nicaragua and Costa Rica, on the Atlantic slope. In the Great Lakes of Nicaragua to the Rio Matina in Costa Rica.

HABITAT: pH: 6.5-8.7, GH: 3-4, KH: 3-4, T: 23-34° C.

DISTINCTIVE CHARACTERISTICS : SL: 185 mm. This species has a lime green body coloration with hundreds of spangles across the fins, body, and lower facial area. This is one of the most beautiful cichlids from Central America, resembling the *Geophagus* species from South America. It has a long, extended snout for digging and scooping the substratum for edible matter. The lips are slightly thickened, suggesting their ability to feed among the larger pebbles and stones of the river's bottom substrate. I have observed a

unique parental behavior with pairs of these fishes in nature and in captivity. Surrounding pairs will breed at about the same time, as if some form of stimulus triggered this reproductive response. They become more tolerant of the adjacent brooding pairs to the point that they will babysit the fry of the other pairs when those fry join with their own spawn. Occasionally all the fry from the surrounding pairs will gather together under the watchful eyes of the parents, forming a community brood center. After a few minutes, the fry will disband and rejoin the tending pairs. I am not sure whether the fry rejoin their respective parents or take up residency with a different set of adults.

C. (A.) tuyrense MEEK & HILDEBRAND, 1913

DISTRIBUTION: Panama, on the Pacific slope. In the Rio Bayano basin.

HABITAT: pH: 7.0-7.9, GH: 3-4, KH: 3-4, T: 25-30° C.

DISTINCTIVE CHARACTERISTICS : SL: 235 mm. This fish is decidedly herbivorous and, in fact, will eat terrestrial plants when they are submerged by the swollen rivers during the rainy season. It is not a very attractive cichlid, as its body coloration is gray to brown.

Cichlasoma (Amphilophus) tuyense. Photo by Hans J. Mayland.

Type species: *Cichlasoma centrarchus* (Gill, 1877)

This small group of small *Cichlasoma* species comprises the dwarf representatives of Central America. Closely related to the subgenus *Theraps*, they have ovate bodies, moderately rounded cranial profiles, and small mouths. Their caudal is truncated or rounded, and they have more spines in the dorsal and anal fins than other North and Central American cichlids. Because they are small in size, they make an easy mouthful to swallow, but their many spines help protect them from various predators. They live and feed in the shallows of the shingle portions, or convex banks, of medium to small sized rivers or along the banks of the lakes they inhabit. They are moderately to strongly sexually dimorphic with respect to size, nuchal hump development, and color pattern, with the female being the more colorful sex. Most females of this subgenus have a black spot or ocellus in the middle of the dorsal fin. The females exhibit a black brood pattern during courtship and generally up until the third or fourth week after spawning. They are generalized feeders of the bottom detritus, consuming most anything edible, including plants and aquatic invertebrates. They spawn in rock or wood cavities and in sand pits or depressions. They make excellent pets for aquarists who prefer smaller fishes or have small aquaria. They are easy to spawn in captivity and are very hardy, making them good selections for advanced and beginner hobbyists alike.

Cichlasoma (Archocentrus) centrarchus. **Rio Zapote, Costa Rica.**

C. (Ar.) centrarchus (GILL & BRANSFORD, 1877)

COMMON NAME: Flier cichlid; the Chorotegas and Niquiranos name is *viejitos*.

DISTRIBUTION: Honduras and Costa Rica, on the Atlantic and Pacific slopes. From the Rio Choluteca in SW Honduras to the Rio Zapote, an effluent of Lago de Nicaragua, in NW Costa Rica. Atlantic slope of central Nicaragua from the Rio Grande to the Rio Matina in Costa Rica. I have been told by natives of the Niquiranos (Mosquito Indians) that this fish inhabits all of the Atlantic coast of Nicaragua.

HABITAT: pH: 6.5-7.2, GH: 1-4, KH: 2-4, T: 26-36° C.

DISTINCTIVE CHARACTERISTICS: SL: 110 mm. This species is found predominately in the shallow waters or swampy areas of lakes and rivers. They often are found in the many eutrophic oxbow lakes, ponds, roadside ditches, and remnant pools of floodplains where the vegetation is thick. They prefer warm temperatures and thrive in stagnant waters with little

problem. They are open water substrate spawners and will lay their eggs in the extreme shallows where predation is minimized. Their fry feed on their parents' dermal mucus for the first few days. Their body coloration is a yellow-green, with light blue or robin's egg blue sides and gill covers. Seven black vertical bars cover the body and extend just a bit into the dorsal fin. The females turn a brilliant blue coloration and their bars become wide and intense. Three

Cichlasoma (Archocentrus) nigrofasciatum. **Female. Rio San Juan, Nicaragua.**

distinct black spots occur behind the eye, two on the opercle, and one at the upper region of the body behind the gill cover.

C. (Ar.) nigrofasciatum (GUENTHER, 1869)

COMMON NAME: Convict; its native Spanish name in Nicaragua is *punto rojo* and in Costa Rica is *punto naranja*.

DISTRIBUTION: Pacific slope of

Guatemala, from the Rio Suchiate to the Rio Grande de Taracoles in NW Costa Rica; Atlantic slope of Honduras, from the Rio Aguan to the Rio Guarumo in Panama.

HABITAT: pH: 6.4-7.6, GH: 3-4, KH: 3-4, T: 20-36° C.

DISTINCTIVE CHARACTERISTICS: SL: 100 mm. This species inhabits flowing water from small creeks and streams to the shallows of the large and fast flowing rivers. They prefer rocky habitats and find sanctuary in the various cracks and crevices provided by this type of environment. They are omnivorous, eating aquatic insects, invertebrates, bottom detritus, algae, seeds, and leaves. They prefer to lay their eggs on the vertical sides or ceilings of small caves and produce about 100-200 offspring per spawning. Their body coloration is charcoal gray with six vertical bars. The various females exhibit different colors on their breasts, undersides, and fins, depending on their locality. Two color morphs I have collected are extremely beautiful. The females from the rivers surrounding Lago de Nicaragua have a brilliant red coloration on their breasts and lower flanks, turquoise anal fins, and yellow and orange dorsal fins. Another yellow-orange color variant exists about 50 kilometers to the south in the Rio Zapote, Costa Rica. The variant from the Rio Sixaola drainage has orange undersides with a metallic blue face and fins. A light pink variant has been produced in captivity.

C. (Ar.) octofasciatum (REGAN, 1903)

SYNONYMS: *Cichlasoma hedricki* Meek, 1904, *Cichlasoma biocellatum* Regan 1909

COMMON NAME: Jack Dempsey

DISTRIBUTION: Atlantic slope, from Rio Coatzacoalcos in SE Mexico to the Rio Sarstoon in Belize.

HABITAT: pH: 7.4-8.0, GH: 13-15, KH: 5-6, T: 26-30° C. Widespread throughout the Rio Usumacinta basin, and

Cichlasoma (Archocentrus) octofasciatum. Photo by Burkhard Kahl.

also found in the Lago Peten basin. This rather robust species prefers the lower and middle sections of the rivers in the slower moving waters. It prefers a soft substrate of sand and mud.

DISTINCTIVE CHARACTERISTICS: SL: 135 mm. This is a beautifully colored cichlid that prefers the coastal plains and slow moving waters of the lower river valleys. The variety from southern Mexico, its northern range limit, has a black body with horizontal rows of turquoise-blue spangles or spots. The variety from the Rio Dandriga, its southern range limit, has a charcoal gray body with approximately 13 horizontal rows of iridescent spots extending from the gill cover to the caudal peduncle. The rows of spots below the lateral line are blue, and the rows above this line are green. The dorsal, anal, and caudal fins show heavy spotting and spangling. The dorsal fin is edged in red, and the breast and lower flanks are bright red. Loiselle says this cichlid originally was placed in the subgenus *Archocentrus* because of its elevated fin spine counts, but this has been questioned by some authorities. Some feel it should be placed in *Nandopsis* for lack of a better alternative, but it probably should be placed in its own monotypic subgenus.

C. (Ar.) sajica BUSSING, 1974

DISTRIBUTION: Pacific slope of Costa Rica, from the Rio Parrita to the Rio Coloradito.

HABITAT: pH: 7.7-8.2, GH: 2-5, KH: 2-6, T: 25-30° C.

DISTINCTIVE CHARACTERISTICS: SL: 80 mm. This species inhabits the smaller rivers and rivulets that have moderate to strong currents, but it is not found in the rapids. It prefers smaller rocks and gravel from which to feed and in which to

Cichlasoma (Archocentrus) sajica. Female. Rio Esquinas, Costa Rica.

breed. It is found up to 2000 feet of elevation. It is omnivorous, consuming algal filaments, aquatic insects, seeds, and bottom detritus. It is a cave or crevice spawner, preferring to attach its eggs to an oblique or vertical surface. They have about 200 offspring which feed on the parents' dermal mucus. The males are of a bright blue coloration with a red-streaked caudal fin. The front half of the dorsal fin is blue, and the rear section is streaked in red. The older males possess a nuchal hump. The females are basically brown-green, and their sides are flecked in pastel pink-orange. The fins are also this pastel coloration. During breeding, the anterior half of the female's body turns black, and the rear half turns white, with a thick, vertical bar through the center of the body and dorsal fin, separating the two contrasting colors.

Cichlasoma (Archocentrus) spilurum. **Rio Polochic, Guatemala.**

***Cichlasoma (Archocentrus) septemfasciatum.* Female. Lago Arenal, Costa Rica.**

C. (Ar.) septemfasciatum REGAN, 1908

DISTRIBUTION: Atlantic slope of Nicaragua from the Rio Grande to the Rio Sixaola basin, the boundary of Costa Rica and Panama. One variant is found in the NW portion of Costa Rica in Laguna de Arenal.

HABITAT: pH: 6.6-7.6, GH: 3-8, KH: 4-8, T: 21-27° C.

DISTINCTIVE CHARACTERISTICS: SL: 100 mm. This species is found in rivers and rivulets of all velocities, slow moving to fast flowing. Their diet consists of algae, leaves, fruits, aquatic insects, and bottom detritus. They are found up to 1800 feet of elevation. They spawn in the caves and crevices of the rocks that landscape the river's bottom and produce about 200-250 offspring per spawning. There are basically three different color variants or color morphs, all with bright green eyes and six vertical bars. All females exhibit a large spot or ocellus in the middle of the dorsal fin. The females of the northern variant have metallic orange sides and light blue abdomens. The males and the females of the southern variant from the Rio Sixaola have horizontal rows of yellow spots on the area below the lateral line. The females have bright yellow sides with a black abdomen and a large black and yellow spot in the middle of their dorsal fin. Both the males and females of the color variant from Laguna de Arenal have a maroon body and maroon fin coloration.

C. (Ar.) spilurum (GUENTHER, 1862)

SYNONYM: *Cichlasoma cutteri* Fowler, 1932

DISTRIBUTION: The Atlantic slope of Belize, from the Belize River to the Rio Aguan in northern Honduras.

HABITAT: pH: 7.3-8.0, GH: 4-9, KH: 3-6, T: 28-32° C.

DISTINCTIVE CHARACTERISTICS: SL: 110 mm. *C. spilurum* inhabits both lacustrine and riverine environments, preferring the shallows and bank areas. This species is found over sand, mud, and rock bottoms, and prefers the slower moving waters of the lower river valleys. They are prolific spawners of the open substrate, usually laying their eggs in depressions in the sand. They produce 300-400 offspring per spawning. Older males have large nuchal humps. The variants in the northern ranges have a light green body with maroon heads and about 7 vertical bars. The race from the Lago de Izabal basin is a much more colorful animal, with a yellow face, breast, and undersides. The dorsal fin is streaked in maroon, and the anal fin is a bright aqua green.

C. (Ar.) spinosissimus (VAILLANT & PELLEGRIN, 1902)

SYNONYM: *Cichlasoma immaculatum* Pellegrin, 1904

DISTRIBUTION: Atlantic slope in Guatemala, from the upper river system of the Rio Polochic.

HABITAT: pH: 8.0, GH: 6, KH: 4, T: 26-28° C.

DISTINCTIVE CHARACTERISTICS: SL: 110 mm. This species differs from *C. spilurum* in that they have a higher number of spines in the anal fin: 11-12. This tan colored fish has numerous black spots covering the body in an interesting irregular pattern.

Type species: *Heros cyanoguttatum* Baird & Girard, 1854

These are medium sized *Cichlasoma* with oval, slightly elongated bodies and moderately rounded or convex cranial profiles. The mouth is small to medium sized and the jaws are slightly protractile. The outer series of teeth are enlarged and slightly rounded. The color pattern is based on serially repeated lateral spots and bars. Breeding coloration is characterized by contrasting light and dark areas and well developed reverse countershading. They are open water pit spawners and basic omnivores, relying essentially on vegetable matter.

C. (H.) bocourti (VAILLANT & PELLEGRIN, 1902)

DISTRIBUTION: Guatemala, on the Atlantic slope. Found in Lago de Izabal and the lower Rio Polochic.

Cichlasoma (Herichthys) bocourti. Illustration by John R. Quinn.

HABITAT: ph: 8.0, GH: 6, KH: 4, T; 26-28° C.

DISTINCTIVE CHARACTERISTICS: Loiselle suggests that this species very well could be of the subgenus *Amphilophus* because of its color pattern and morphology. Dentition and biogeographic grounds are the only evidence for placing this species in the subgenus *Herichthys*.

C. (H.) carpinte (JORDAN & SNYDER, 1899)

DISTRIBUTION: Northeast Mexico, on the Atlantic slope. Widespread in the Rio Panuco and Rio Soto la Marina basins.

HABITAT: pH: 6.7-7.8, GH: 6-24, KH: 6-10, T: 23-33° C.

DISTINCTIVE CHARACTERISTICS: SL: 170 mm. This species is similar to *C. cyanoguttatum* and is its replacement species in the Rio Panuco basin. This species has less spotting, but of larger size and usually of a

Cichlasoma (Herichthys) carpinte. Laguna del Chairel, Mexico.

Cichlasoma (Herichthys) cyanoguttatum.
Photo by Hejins.

turquoise green color. The base coloration ranges from gray to brown to maroon. Many color variants exist, and more scientific work must be completed to fully understand this complex.

C. (H.) cyanoguttatum (BAIRD & GIRARD, 1854)

SYNONYMS: *Cichlasoma pavonaceum* Garman, 1881, *Cichlasoma laurae* Regan, 1908.

COMMON NAME: Texas cichlid

DISTRIBUTION: Northeast Mexico and Texas, on the Atlantic slope, from the Rio Conchos to the Rio Grande basin.

HABITAT: pH: 6.7-7.8, GH: 6-54, KH: 6-9, T: 20-33° C.

DISTINCTIVE CHARACTERISTICS: SL: 180 mm. This is the northernmost member of the genus *Cichlasoma* and the only member native to the U.S. (southern part of Texas). We have found them in Lake Travis, just north of Austin, Texas. A population exists in the Alafia River, Tampa, Florida, introduced by widespread flooding of the resident fish farmers. This widespread species is moderately deep bodied and presents a series of 5 to 6 lateral blotches, the first and most prominent located on the mid-flank.

These blotches usually occur with faint crossbars and a dark spot at the base of the caudal fin. The body coloration is light gray with numerous pale blue spots arranged irregularly throughout the body and fins.

C. (H.) sp. "Ebano" (1991)

DISTRIBUTION: Northeast Mexico, Atlantic slope. Found in an earthen agricultural canal, Canal Ebano Principal.

HABITAT: pH: 7.2, GH: 6, KH: 7, T: 28-30° C. Found in the lower river valleys in the savannahs near Tampico and prefers a soft or sandy bottom.

DISTINCTIVE CHARACTERISTICS: SL: 230 mm. The dentition and morphological structure of this species are similar to those species of the subgenus *Herichthys*. It grows quite large, which would make this species the largest member of this subgenus. Its structure resembles that of *C. fenestratum*, the northernmost species of *Theraps*. However,

no *Theraps* species has been recorded from the Panuco basin. Only scientific research will determine the status of this undescribed species. As a juvenile it has a pale olive-yellow coloration with 6-7 irregular spots or blotches along the mid-body line. Vertical bands or bars extend from the spots to the base of the dorsal fin. Adults develop an orange-yellow coloration, with each scale bordered in light blue. Blue spangling and wavy streaks appear on the face, forehead, and operculum, as well as the dorsal and caudal fins. The anal, pelvic,

Cichlasoma (Herichthys sp. "Ebano". Ebano Canal, Mexico.

and ventral fins are a light yellow color with an orange cast.

C. (H.) labridens (PELLEGRIN, 1903)

DISTRIBUTION: Northeast Mexico, on the Atlantic slope. Endemic to the Rio Verde and the Laguna Media Luna system in the Rio Panuco basin.

HABITAT: pH: 6.7-7.2, GH: 43-47, KH: 9-10, T: 26-30° C.

Cichlasoma (Herichthys) labridens. Laguna de la Media Luna, Mexico.

DISTINCTIVE CHARACTERISTICS: 160 mm. *C. labridens* is one of the most colorful of the Panuco cichlids, with a bright yellow dress. A blue-black area extends from the underside of the head, below the level of the nostril and bottom of the orbit, as a rectangular blotch over the operculum, pelvic and pectoral fins, breast, and anterior flank. Another blotch occurs on the caudal peduncle and posterior flank, sometimes extending into the anal fin.

C. (H.) sp. "Lahillas" or "labridens-blue" (1989)

DISTRIBUTION: Northeast Mexico, Atlantic slope. Found in the Las Lahillas reservoir of the northern Rio Panuco basin.

HABITAT: pH: 7.6, GH: 8, KH: 6, T: 25-30° C.

DISTINCTIVE CHARACTERISTICS: SL: 180 mm. This is a deep-bodied species that has dentition and other morphological similarities to the subgenus *Herichthys*. The gut contents were heavy with vegetable matter and snail remains. I caught this cichlid in a man-made reservoir named Las Lahillas, or Flagstone Dam. It is found sympatrically with a *C. carpinte* variant, but it was not nearly as common, less than 1 per 50. The juvenile color pattern is an olive green with a broad black area along the mid-body line from behind the eye to the base of the caudal peduncle. About at the middle of this line, the black extends upwards through the back and into the dorsal fin. Five to six faded vertical stripes are located behind this prominent black area, and black specks appear in the dorsal and caudal fins. The area below this mid-line and the anal fin is a sky-blue or robin's egg blue color. In both sexes a thin red stripe extends across a charcoal colored forehead from the top of one opercle to the top of the other.

C. (H.) sp. "Altamira" (1989)

DISTRIBUTION: Northeast Mexico, Atlantic slope. Found in Laguna de Altamira, a brackish lagoon north of Tampico in Altamira, Tamaulipas.

Cichlasoma (Herichthys) sp. "Altamira". Laguna de Altamira, Mexico.

Cichlasoma (Herichthys) sp. "Lahillas". Las Lahillas, Mexico.

HABITAT: pH: 7.8, DH: 22, KH: 9-10, T: 28-30° C.

DISTINCTIVE CHARACTERISTICS: SL: 185 mm. This slightly elongated species has 6-7 spots along the mid-body line and an olive green body color. The scales are bordered in a light blue color from the ventral surface to an area just above the mid-body line. Light blue streaks are found on the face and the operculum. Blue streaking is found in the dorsal, caudal, anal, and pelvic fins.

C. (H.) pearsei (HUBBS, 1936)

DISTRIBUTION: Southeast Mexico and northern Guatemala in the Rio Usumacinta basin and the Rio Tulija system of the Rio Grijalva basin.

HABITAT: pH: 7.0-7.8, GH: 3-9, KH: 3-8, T: 26-30° C. Typical lacustrine and lower river valley species with slight tolerance of brackish water or the estuarine environment.

DISTINCTIVE CHARACTERISTICS: SL: 200 mm. This deep bodied species is one of the largest members of the subgenus *Herichthys*. Its bright yellow body contrasts with a maroon-black belly and abdomen. The scale bases in this maroon-black color are bordered by bright yellow, forming a reticulated diamond pattern. A single spot occurs at the base of the caudal fin. The dorsal and caudal fins are a charcoal gray with light blue spangling. The anal and pelvic fins are an amber-orange. This "leaf chopper" prefers both the aquatic and terrestrial plants that line the river banks. It takes several days to clean out the gut content, indicating a long intestinal tract.

Cichlasoma (Herichthys) pearsei. Rio Tulija, Mexico.

Nandopsis

Type species: *Acara adspersa* Guenther, 1862

Regan (1905) created the subgenus *Parapetenia* to include species of *Cichlasoma* with enlarged anterior canines. Kullander (1983) noted that *Nandopsis* Gill (1862) has priority over *Parapetenia* Regan (1905) because the type species of each is the same. This *Cichlasoma* group is comprised of the specialized piscivores that are the predators of the American cichlid communities. They are basically lacustrine, but most will venture into the slow moving waters of the lower river valleys. They are medium to large sized and have wide mouths with strongly protractile jaws. The few teeth in the upper and lower jaws are enlarged anteriorly. The jaws are well equipped with strongly developed pseudocanines, one pair centered in the upper jaw and two pairs in the lower. The caudal is subtruncated or rounded, and the bases of the soft dorsal, anal, and caudal fins are scaled. Adult males have serially repeated spots that form a pattern over a colored body. Females have large serially repeated spots along the mid-body line, and their coloration is greatly enhanced while sexually or parentally active, but they lack the overall mottled pattern. They prefer to lay their eggs in large tunnel caves hollowed out along the lakes' edges or underneath the rivers' banks. The size of the cave is modified to their specifications by digging and chewing away at the hardened substrate until the dimensions are about 1 $\frac{1}{2}$-2 times their body length in circumference. The males and females clean away the sand and other loose debris and deposit it just outside the cave's entrance, forming a kind of beach or front porch. The female deposits from 300-2000 eggs, depending on the species, on the interior vertical walls. During the prespawning and postspawning periods, the male remains outside the entrance and sets up a 2-4 cubic meter territory. He is on constant vigil and will challenge anything that invades his chosen boundaries or parameters. Only occasionally will the male and female exchange places and assume each other's role. After spawning, the pair diligently watch over their young, leading them up and down the lake's shoreline or slow moving river banks in constant search of food. For the first few weeks they tend their broods in the upper sections of the water, rarely submerging deeper than 2 meters. The male spends less and less time with the group after the first couple of weeks until they eventually leave for good. The female remains on guard for several weeks until the juveniles are capable of surviving on their own. This is when they are about 5-6 weeks of age, or about 2-3 inches in length. Needless to say, predation is high in the first few weeks on their own.

C. (N.) bartoni (BEAN, 1892)

DISTRIBUTION: Endemic to the upper Rio Verde and the Laguna de la Media Luna system of the Rio Panuco basin in the State of San Luis Potosi, Mexico.

HABITAT: pH: 6.6, GH: 4-7, KH: 9, T: 26-30 C. This fish

Cichlasoma (Nandopsis) bartoni. Laguna de la Media Luna, Mexico.

occurs sympatrically with *Cichlasoma labridens.* They flourish best in the region of Laguna Media Luna, a large marsh lagoon fed by warm springs, where the water is very clear, bluish, and has a strong sulfur odor. Unfortunately, the surrounding community is agriculturally oriented, and the runoff of the various pesticides and herbicides is polluting this small, unique system. Many of the fishes are weakened with various external diseases as a result, and only time will tell what happens to this fragile ecosystem. Besides dining on other smaller fishes, this species has added invertebrates to its diet of algae and aquatic insects.

DISTINCTIVE CHARACTERISTICS: SL: 180 mm. This species represents the smallest member of the subgenus *Nandopsis.* Adults display an aquamarine coloration with a wide, blackened pattern occurring above and below the lateral line. This irregular, blackened pattern is common

to all members of the Panuco fluvial system except for *Cichlasoma carpinte* and *C. cyanoguttatum.* However, when in breeding condition, both sexes develop contrasting patterns of white above the lateral line and blue-black below the lateral line. Both color patterns are uniform and uninterrupted. Their spawns are small in size, numbering between 200-300.

C. (N.) beani (JORDAN, 1888)

DISTRIBUTION: Pacific slope of Mexico, from the Rio Presidio to the Rio Grande de Santiago.

HABITAT: pH: 7.0-7.2, GH: 3-4, KH: 5-6, T: 24° C.

DISTINCTIVE CHARACTERISTICS: SL: 300 mm. This fish is found in the lower river valley sections and is the northernmost cichlid representative on the Pacific slope. It has 9 elongated blotches that create an almost perfect horizontal line from behind the pectoral fins to the caudal peduncle. It has a similar pattern when breeding to most of the cichlids found north of the

Cichlasoma (Nandopsis) beani.

Colima mountain chain. The scales of the body are formed in a reticulated, honeycomb pattern. Two black stripes cross the head just above the eyes. There is a black stripe along the dorsal fin.

C. (N.) dovii (GUENTHER, 1864)

COMMON NAME: Wolf cichlid; its Spanish or native name is *guapote blanco*, and its native Indian name is *laguneros*, meaning large fish.

DISTRIBUTIONS: Atlantic slope of Honduras, from the Rio Aguan to the Rio Moin in Costa Rica. Pacific slope of Honduras, from the Rio Yeguare to the Rio Bebedero in Costa Rica.

HABITAT: pH: 7.5-8.8, GH: 2-5, KH: 2-4.5, T: 26-33° C.

DISTINCTIVE CHARACTERISTICS: SL: 500 mm. This species is lacustrine in nature, but also inhabits the various lower and middle river valleys. Hence it is widespread on both slopes. They are avid cavern diggers, tunneling into the upper regions of the shoreline along the lakes and rivers in which they reside. This species is the largest of all the *Cichlasoma* species, reaching lengths over 500 mm. This is one of the predominant food fishes in the marketplace of Grenada on the northeastern shore of Lago de Nicaragua and is eaten baked or as fillets. Being piscivorous, this species is at the top of the trophic pyramid and therefore is not abundant. They are always on the hunt for smaller fishes on which to feed. Their torpedo body shape allows them to burst through the water with great speed, enabling them to suck in their quarry with their highly protractile jaws. Their pseudocanines can firmly hold onto the bigger prey that are too large to swallow. The body coloration of the male is light green, with an overlying rose or purple cast, and the fins are metallic turquoise. The males have about 14 horizontal rows of black spots forming an attractive serial pattern.

These spots continue into the gill cover, across the facial area under the eye, and into all the fins. The males develop a knot on their forehead rather than a nuchal hump. The females have a lime green to yellow coloration with a wide spotted or broken lateral line. This species has spawns numbering about 1000-1500. There is a golden form in the Rio Tortuguero in NE Costa Rica. There is also an orange-red morph identified by Bussing in the Rio Puerto Viego de Serapiqui in Costa Rica. I have seen them in the river, but they live in very deep water and have eluded my attempts to capture them.

C. (N.) friedrichsthalii (HECKEL, 1840)

SYNONYMS: *Cichlasoma multifasciatum* Regan, 1905

DISTRIBUTION: Atlantic slope, from Rio Coatzacoalcos basin in Mexico to the Rio Sarstoon in Belize. The most beautiful yellow race is found in Lago Peten in northern Guatemala.

HABITAT: pH: 7.4-8.0, GH: 13-15, KH: 5-6, T: 26-30° C.

DISTINCTIVE CHARACTERISTICS: SL: 280 mm. This species is both lacustrine and riverine, preferring the slower moving waters. They feed primarily upon other live fishes by remaining motionless and allowing the victim to come within close striking distance or by slowly stalking them and overtaking them with a sudden burst of speed. The adult males have a yellow coloration below the lateral line, which extends into the caudal, anal, and pelvic fins. They display a bronze to orange coloration above the lateral line, which extends through the gill cover and onto the cheeks and

Cichlasoma (Nandopsis) dovii. **Lago de Nicaragua, Nicaragua.**

Cichlasoma (Nandopsis) friedrichsthalii. Belize River, Belize.

Cichlasoma (Nandopsis) friedrichsthalii. Belize River, Belize.

throughout the dorsal fin. The body is mottled in adult males, with each scale being edged in black below the lateral line. Serial rows of spots adorn the upper body above the lateral line and throughout all the fins, giving this fish an attractive pattern. The females display a deep yellow coloration with a broken row of irregular spots below the spotted mid-body line. Both sexes have seven to nine vertical bars and two large and irregular black spots on the gill cover and one beneath the eye. Their spawns number about 500-700 eggs.

C. (N.) grammodes TAYLOR & MILLER, 1980

SYNONYMS: *Cichlasoma* sp. Miller, 1966, *Cichlasoma mento* Miller, 1976

DISTRIBUTION: This species is known only on the Pacific slope of Mexico in the Rio Grande de Chiapa basin, from Villa Flores, Chiapas, Mexico, to the Rio Lagartero, Huehuetenango, in extreme western Guatemala.

HABITAT: pH: 7.5-8.0, DH: 10-20, KH: 12-18, T: 24-30° C. This species exists in the lower to middle river valleys, as well as the lake, Presa de Angostura. It can be found over rocks, sand, silt, or mud.

It inhabits most all the habitats available within its small territorial boundaries.

DISTINCTIVE CHARACTERISTICS: SL: 180 mm. They like the moderate flowing streams and large rivers with strong currents which are clear, clean, and high in oxygen content. They prefer rocky habitats with gravel, boulders, marl, and sand bottoms. Their bodies are elongated, with extremely long heads and large mouths. The lower jaw protrudes slightly past the upper jaw, and the lips are strong and massive. The preorbital region is rather olive green. The color pattern

Cichlasoma (Nandopsis) friedrichsthalii. Laguna Cahabon, Guatemala.

Cichlasoma (Nandopsis) friedrichsthalii. Female. Belize River, Belize.

Cichlasoma (Nandopsis) grammodes. Rio Grande Chiapas, Mexico.

is much less pronounced. The adult males have elongated dorsal and anal fins and a subtruncate caudal. Juveniles have an interrupted lateral bar from the operculum to the caudal peduncle. Except for the back and dorsal base, the scales are turquoise-green and bordered with maroon red margins, forming a reticulated diamond pattern. The lower face, gill cover, and the area above and behind the gill cover are spangled. Except for the caudal fin, the fins are heavily spangled and streaked

Cichlasoma (Nandopsis) istlanum. Rio Armeria, Mexico.

consists of a broad, sometimes indistinct, horizontal stripe extending from the gill cover to the caudal. The lower half of the body below the lateral stripe is an aquamarine color covered with maroon spots arranged in rows along the scales. A series of seven thin, maroon lines occur on the side and top of the head, extending from the eye to the tip of the snout. The upper half of the body above the lateral stripe is a mauve color with the same symmetrical maroon spotting. The dorsal, caudal, and anal fins are bright blue and heavily spotted.

C. (N.) haitiensis TEE-VAN, 1935
DISTRIBUTION: The West Indies island of Hispaniola.

C. (N.) hogaboomorum CARR & GIOVANNOLI, 1950
DISTRIBUTION: Pacific slope of Honduras in the lower part of the Rio Choluteca.
DISTINCTIVE CHARACTERISTICS: It

resembles *Cichlasoma urophthalmus.*

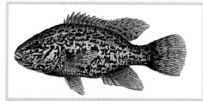

Cichlasoma (Nandopsis) haitiensis. Illustration by John R. Quinn.

C. (N.) istlanum (JORDAN & SNYDER, 1899)
SYNONYMS: *Heros mento* Vaillant and Pellegrin, 1902
DISTRIBUTION: Pacific slope of Mexico from the Rio Armeria to the Rio Papagallo.
HABITAT: pH: 7.0-7.4, GH: 3-5, KH: 3-5, T: 24-28° C.
DISTINCTIVE CHARACTERISTICS: SL: 140 mm. This species is similar to *Cichlasoma grammodes* in color and pattern but is smaller and has a less massive cranial profile. Its snout is blunt and rounded, and the lip structure

in turquoise-green. During the breeding season the males develop massive nuchal humps. The females are much less intense in coloration.

C. (N.) loisellei BUSSING, 1989
COMMON NAME: *guapote amarillo*
DISTRIBUTION: Atlantic slope of Guatemala in Lago de

Cichlasoma (Nandopsis) hogaboomorum. Illustration by John R. Quinn.

Izabal and the Rio Dulce to the coastal drainages of Laguna de Chiriqui in Panama. Pacific slope of Nicaragua in the Rio San Juan basin, inclusive of the Great Lakes of Nicaragua.

HABITAT: pH: 7.5-8.8, GH: 2-5, KH: 2-4.5, T: 24-37 °C.

DISTINCTIVE CHARACTERISTICS: SL: 190 mm. *C. loisillei* is known from suitable lowland and coastal habitats, and although it is found sympatric

nate. A series of black quadrangles form a prominent discontinuous band from the eye to the base of the caudal fin. Two other conspicuous black markings form a broken band from eye to lower opercle. Also, adults show no vertical barring as that shown by *C. friedrichsthalii*.

C. (N.) managuense (GUENTHER, 1869)

COMMON NAME: Jaguar

Cichlasoma (Nandopsis) managuense. Lago de Nicaragua, Nicaragua.

Cichlasoma (Nandopsis) loisellei. Female. Rio Sixaola, Costa Rica/Panama.

with *C. dovii* and *C. managuense*, it maintains separate territories, choosing the sluggish water of streams, backwaters, and swamps. This guapote is somewhat less piscivorous than the others, preferring aquatic and terrestrial insects and only occasionally taking fish. Their body is moderately deep, and the snout is short, with the mouth turned upward. The lower jaw projects slightly in advance of the premaxillaries. The body coloration is similar to that of *C. friedrichsthalii*, but the configuration of spots is absent and, instead, series of square blotches predomi-

cichlid or Aztec cichlid

DISTRIBUTION: Atlantic slope from Rio Ulua in Honduras to the Rio Matina in Costa Rica. Pacific slope of Nicaragua in the crater lakes and the Great Lakes of Nicaragua. I have collected this species also in the Rio Motagua near Morales on the Atlantic slope.

HABITAT: pH: 7.5-8.8, GH: 2-5, KH: 2-4.5, T: 25-36° C.

DISTINCTIVE CHARACTERISTICS: SL: 250 mm. This species is lacustrine, preferring turbid waters and mud floors of the highly eutrophic lakes, where they feed on small fishes and macroinvertebrates. This robust piscivore is the second

largest species in the genus *Cichlasoma*. Their bodies are basically white with a light purple cast and a black mottled pattern of irregular and connecting spots and wavy lines. They closely resemble the much smaller *C. tetracanthus* from Cuba.

C. (N.) minckleyi KORNFIELD & TAYLOR, 1983

DISTRIBUTION: Atlantic slope of northern Mexico, this species is endemic to Cuatro Cienegas.

HABITAT: pH: 7.6, GH: 54, KH: 9, T: 23-25° C.

DISTINCTIVE CHARACTERISTICS: SL: 175 MM. This species is the northernmost member of the subgenus *Nandopsis* and includes in its diet snails and small fishes. It is similar to *C. cyanoguttatum*, the only cichlid it shares its habitat with, but has the typical snout and dentition of the primitive piscivorous group.

C. (N.) motaguense (GUENTHER, 1866)

DISTRIBUTION: Atlantic slope of Guatemala in the Rio Motagua basin. Pacific slope of Guatemala from the Rio Naranjo to the Rio Choluteca in Honduras.

HABITAT: pH: 7.0-7.4, GH: 5-6, KH: 5-6, T: 26-28° C.

DISTINCTIVE CHARACTERISTICS: 260 mm. This species is both

Cichlasoma (Nandopsis) minckleyi. Photo by Hans-Joachim Mayland.

lacustrine and riverine, but prefers the moderate to fast flowing waters of the lower and middle river valley sections. Its body shape is similar to that of its sibling species, *C. friedrichsthalii* and *C. loisellei*. The smaller, streamlined bodies of these medium sized *Nandopsis* members enable them to maneuver more easily in the current or flow of the riverine biotope. Their thick, powerful bodies give them the ability to chase down their prey with great bursts of speed. They are well equipped to seize and hold on to their victims, if too large to swallow, with their enlarged

Cichlasoma (Nandopsis) motaguense. Rio Motagua, Guatemala.

pseudocanines. After their victim is dead or too weak to escape, they will spit it out and suck it back into their mouths using their frontal canine teeth to crush, tear, and chew away small pieces that then can be swallowed. They also relish the aquatic and terrestrial insects that are abundant in this area of the world. Adult males have

Cichlasoma (Nandopsis) sp. cf. minckleyi.

the usual broken lateral stripe and mottled body and fin pattern. The color pattern of *C. motaguense* is composed of small rust colored square blotches linked together in serial horizontal rows over a pale green body. The females have a brilliant orange to orange-red color, which is intensified in their lower face, gill cover, breast, lower flanks, and anal fin. They exhibit a second horizontal and irregularly shaped stripe just below the lateral stripe, similar to the female *friedrichsthalii*. In addition, the upper back below the dorsal fin is blotched in a charcoal coloration.

C. (N.) sp. cf. *minckleyi*
    NATIVE NAME: *guapota*
    DISTRIBUTION: Endemic to Laguna de las Chorreras, Canal de las Chorreras, and Laguna del Chairel, near Tampico, on the Atlantic slope.
    HABITAT: pH: 7.0-7.2, GH: 20-24, KH: 8-9, T: 25-32° C.
    DISTINCTIVE CHARACTERISTICS: Body orange, with blue spots

and blue line below the eye and on the chin. A few black specks cross the body. This form shares its habitat with *C. carpinte*, *C. pantostictum*, and *C. sp. cf. pantostictum*, but has the typical snout and dentition of the primitive piscivorous group.

C. (N.) *pantostictum* TAYLOR & MILLER, 1983
    DISTRIBUTION: Atlantic slope of northern Mexico from the Rio Sabinas to the coastal Laguna Tamiahua. They seem to be concentrated in the cooler and clearer water of the Laguna Escondida, which is connected with the larger Laguna Chairel.

Cichlasoma (Nandopsis) pantostictum.

HABITAT: pH: 7.0-7.2, GH: 20-24, KH: 8-9, T: 25-32° C.

DISTINCTIVE CHARACTERISTICS: SL: 180 mm. This species is a medium sized *Nandopsis*, about 6-8 inches, that prefers the coastal lacustrine environment, which has a slight salt content, is slightly turbid or eutrophic, and has a mud or sand bottom. They are basically aquatic and terrestrial insect eaters, but will take small fishes when easily available. The fishermen informed us that swarms of small flies gather on the marsh and floating weed vegetation in the late evening during the summer. I was lucky enough to witness this event where the *C. pantostictum* concentrated in large numbers, jumping out of the water to feed upon the masses of flies and their larvae. The adults spawn among the exposed roots at the lagoon edges, and their eggs number about 400-500. The males have a chocolate colored body, and the head and fins are profusely speck-led with small dark brown spots. A diagnostic pattern is developed in breeding adults. The females are an overall olive brown with less speck-ling.

C. (N.) sp. cf. *pantostictum*

SYNONYMS: *Cichlasoma* "lamadridi".

NATIVE NAME: *guapota*
DISTRIBUTION: Endemic to Laguna de las Chorreras, Canal de las Chorreras, and Laguna del Chairel, near Tampico, on the Atlantic slope.

HABITAT: pH: 7.0-7.2, GH: 20-24, KH: 8-9, T: 25-32° C.

DISTINCTIVE CHARACTERISTICS: SL: 150 mm. This rare cichlid is rather deep bodied when compared with the other members of the subgenus *Nandopsis*. The body scales are blue and surrounded or edged with a black border, while a variegated black pattern extends across the body. Only the upper head and shoulder area remain free of this "plaid" pattern, but these areas are heavily speck-led with numerous black dots. The red-tipped dorsal fin also is covered with these black spots, as well as a few larger sized red "egg" spots. The juveniles of this species have prolonged snouts and a black lateral band.

C. (N.) *ramsdeni* FOWLER, 1938

DISTRIBUTION: The eastern side of the mountain ridge on the far eastern side of Cuba.

HABITAT: pH: 6.8-7.2, GH: 3-

Cichlasoma (Nandopsis) sp. cf. pantostictum. Laguna del Chariel, Mexico.

Cichlsoma ramsdeni. Photo by H.R. Axelrod.

4, KH: 3-4, T: 24-30° C.

DISTINCTIVE CHARACTERISTICS: This species is very close to its sibling species, *C. tetracanthus*. Eastern and western Cuba were separated and united about 10 million years ago. It is speculated that this species developed from *C. tetracanthus* during its long geographical isolation.

C. (N.) salvini (GUENTHER, 1862)

SYNONYMS: *Heros triagramma* Steindachner, 1864, *Cichlasoma tenue* Meek, 1906

COMMON NAME: Tricolor cichlid

NATIVE NAME: *peine, mango pinto*

DISTRIBUTION: Atlantic slope of Mexico from the Rio Papaloapan to the Sulphur River near Puerto Barrios, Guatemala.

HABITAT: pH: 7.4-8.0, GH: 13-15, KH: 5-6, T: 26-32° C.

DISTINCTIVE CHARACTERISTICS: SL: 220 mm. This species prefers the riverine biotope and is a member of the smaller sized *Nandopsis*. They

have an extremely pointed cranial profile to accommodate their preference for the moderate to fast flowing waters of the lower and middle river valleys, where they live among the submersed trees and their roots. They feed among the various weeds and aquatic plant life where the macroinvertebrates and other small fishes find sanctuary.

The males have bright green spangling covering the entire body and fins, and the scales from the lower flanks are edged in red. The females are the more beautiful of the sexes in their bright yellow and red coloration. A dark green broken bar runs above and parallel to the lateral stripe and is laced with turquoise spangling. They produce about 500-600 eggs per spawning.

C. (N.) steindachneri (JORDAN & SNYDER, 1899)

DISTRIBUTION: Atlantic slope of northern Mexico. Endemic to the Rio Tamasopo, Rio Gallinas, and Rio Ojo Frio.

HABITAT: pH: 6.7-7.2, GH: 24-47, KH: 8-10, T: 24-29° C.

DISTINCTIVE CHARACTERISTICS: SL: 190 mm. This cichlid is very rare in nature, occurring sympatrically with the currently misidentified black and white color form of *C. labridens*. The type of *labridens* was described from the Rio Verde system and is very different from this fish. *C. steindachneri* is slender-

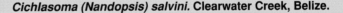

Cichlasoma (Nandopsis) salvini. Clearwater Creek, Belize.

Cichlasoma (Nandopsis) steindachneri. Rio Agua Buena, Mexico.

bodied, with a long, pointed head. This fish is a snail crusher as well as a predator of smaller fishes. Its mouth is large and oblique, with the lower jaw projecting. Adults have yellow bodies and heads with creamy-white lower flanks and the typical black diagnostic pattern as most of the Panuco cichlids. As small specimens, they are difficult to differentiate from the black and white *labridens*-like species found here. Only when they are larger, about 3", is it possible to tell them apart. This has led to great confusion for the field workers and scientists.

C. (N.) tetracanthus (VALENCIENNES, 1831)

SYNONYMS: *Acara adspersa* Guenther, 1862, *Acara cubensis* Steindachner, 1863

COMMON NAME: Cuban cichlid

DISTRIBUTION: The island of Cuba.

HABITAT: pH: 6.8-7.2, GH: 3-4, KH: 3-4, T: 24-30° C.

DISTINCTIVE CHARACTERISTICS: SL: 240 mm. An island mem-

ber of the subgenus *Nandopsis*, this primitive species is not that specialized, as it has little competition for food. Its contrasting and alternating pattern of black irregular blotches, larger toward the front of the fish and smaller toward the rear, on a bright white body makes this fish desirable for ornamental use. Some populations

have a rosy-violet cast. Their spawns average about 300.

C. (N.) trimaculatum (GUENTHER, 1869)

SYNONYMS: *Cichlasoma mojarra* Meek, 1904, *Cichlasoma evermanni* Meek, 1904, *Cichlasoma centrale* Meek, 1906, *Cichlasoma cajali* Alvarez & Gutierrez, 1952

COMMON NAME: Red eyed cichlid, three spot cichlid

DISTRIBUTION: Pacific slope of Mexico in the Laguna Coyuca to the Rio Lempa in El Salvador.

HABITAT: pH: 6.4-7.0, GH: 4-5, KH: 4-6, T: 26-30° C.

DISTINCTIVE CHARACTERISTICS: SL: 365 mm. This species is found in the slow moving waters of the lower river valleys and prefers mud and sand bottoms, where it lives in the roots and weeds. Its diet consists of small fishes, macroinvertebrates, and aquatic and terrestrial insects. This species is deep bodied and has a cranial profile similar to that of the *Amphilophus*. Males have a

Cichlasoma (Nandopsis) tetracanthus? Cuba.

Cichlasoma (Nandopsis) trimaculatum. **Photo by Stawikowski.**

strong three spot or blotch pattern, one of which is situated in the region of the upper shoulder area. They are predominately green, with a yellow head and yellow undersides and a bright claret colored throat. The females turn a dark blue-black coloration during periods of sexual activity. Large females can lay more than 1000 eggs.

C. (N.) umbriferum MEEK & HILDEBRAND, 1913

COMMON NAME: Turquoise cichlid

DISTRIBUTION: Pacific slope of Panama in the Rio Tuira basin to the Rio Magdalena basin in Colombia on the Atlantic slope.

HABITAT: pH: 7.2-8.1, GH: 3-5, KH: 3-5, T: 27-32° C.

DISTINCTIVE CHARACTERISTICS:

SL: 475 mm. This robust species is the southernmost representative of the subgenus *Nandopsis* and competes with *C. dovii* for the largest *Cichlasoma* award. They are open water swimmers, and their primary diet is other live fishes. Their mouth is quite large and protractile, capable of swallowing 3-inch-long fishes. They are quite fond of the prolific *Astyanax* from this area, and it is a common sight to see a few specimens stalking and chasing a school of these characins. Guapotes with protrusive lower jaws are well equipped to chase and capture upper water level fishes, such as species of *Astyanax* and *Poecilia*. Their olive-yellow body is entirely covered with metallic blue spots and two large black blotches are found, one central on the body, the other on the tail base.

C. (H.) sp. "Tamasopo"

DISTRIBUTION: Northeast Mexico, on the Atlantic slope. Endemic to the Rio Tamasopo system of the Rio Panuco basin.

HABITAT: pH: 6.7-7.2, GH: 24-47, KH: 8-10, T: 24-29° C.

DISTINCTIVE CHARACTERISTICS: SL: 165 mm. Erroneously referred to as *C. labridens*, this species still is undescribed, but is similar to *C. bartoni* and *C. pantostictum*. It probably belongs to the subgenus *Herichthys*. This species is found sympatric with *C. steindachneri*. According to Miller, their extreme similarity in appearance as juveniles and adolescents led to the confusion of these two species. About 1 in every 100 juveniles of this blotchy pattern turned out to be *C. steindachneri*, while the bulk was this undescribed species. As adults, this animal has display colors similar to *C. bartoni*, except that *C. bartoni* has an evenly divided black-white contrasty pattern, while *C.* sp. "Tamasopo" has a disrupted black-white contrasty pattern. This species's body is jet black, except for a white splotchy area on the belly and abdomen and white upper face and upper shoulder area. This latter area is well defined, as the line of demarcation is from the nostrils through the bottom curvature of the eye, across the upper operculum, and to a point where the posterior end of the lateral line ends. The dividing line turns upward through the dorsal. In some populations the posterior end of the upper back may have the same white blotchiness as found on the belly and abdomen. The white forehead is freckled with very small black specks. The dorsal, caudal, and anal fins are streaked in red and blue, and the pelvic fins are dark black.

Cichlasoma (Nandopsis) umbriferum. **Rio Lutera, Panama.**

Cichlasoma (Herichthys) sp. "Tamasopo". Rio Tamasopo, Mexico.

C. (N.) urophthalmus (GUENTHER, 1862)

SYNONYMS: *Heros troscheli* Steindachner, 1867

COMMON NAME: Orange tiger

NATIVE NAME: *castarrica*

DISTRIBUTION: Atlantic slope of Middle America, from the Rio Coatzacoalcos basin southward into Nicaragua, including the Yucatan Peninsula and Isla Mujeres.

HABITAT: pH: 7.4-8.0, GH: 13-15, KH: 5-6, T: 26-30° C.

DISTINCTIVE CHARACTERISTICS: SL: 280 mm. This species prefers the coastal lagoons and rivers and will tolerate marine conditions. I have seen them in the mangrove swamps of St. George's Cay, about 3 miles offshore of Belize City. I also have seen them on the northern rocky coast of Isla Mujeres, about 1 mile off the coast of Cancun. They are piscivorous and work the roots and weeds for macroinvertebrates and small fishes. Their orange colored bodies have 5-6 margined vertical dark bars and a single prominent dark spot on the upper caudal fin base.

C. (N.) vombergae LADIGES, 1938

DISTRIBUTION: The Dominican Republic, Hispaniola. N.B. This species often has been considered a synonym of *C. (N.) haitiensis*. According to Konings (1989), it is well separated from that species geographically by the Central Highland. It may now be extinct, probably due in large part to human intervention.

Cichlasoma (Nandopsis) sp. "Tamasopo". Rio Tamasopo, Mexico.

Theraps

Type species: *Theraps irregularis* Guenther, 1862.

THERAPS: GROUP 1

Theraps can be split into two groups: (1) larger, deeper bodied lacustrine and moderately slow-flowing riverine species and (2) smaller, more elongated faster current riverine species. Group 1 can be subdivided into two categories, the first of which includes the lacustrine species.

The lacustrine species prefer mesotrophic and eutrophic lakes characterized by slightly cloudy to turbid waters, sediment buildup, and large numbers of aquatic plants. They feed on the bottom detritus of vegetable matter, aquatic and terrestrial plants, and seeds and fruits that drop from overhanging trees. The upper profile of the snout is usually slightly convex and the terminal mouth is moder-

ate in width. The cleft of the mouth is small and nearly horizontal and entirely below the level of the eye. The jaws are equal anteriorly or the lower is somewhat shorter. The dorsal fin has 10-15 rays and the anal has 7-12 rays. The pectoral is usually short and only in one species, *C. nicaraguense*, does it extend beyond the origin of the anal; the caudal is either rounded, truncate, or emarginate with

Cichlasoma (Nandopsis) urophthalmus. Sibun river, Belize.

rounded lobes. During periods of sexual activity a pair will mark off a 2-4 meter territory in the shallows near the lake's edge and defend it fearlessly. The pair will dig or fan away the loose sand and mud, forming a depression or pit in the earthen substrate. The female will lay between 500-1000 large eggs and continually fan water across them, thus avoiding possible contamination. Under aquarium conditions, the pair will spawn on a smooth, flat rock because of the uneven texture of the bottom gravel.

The second category of Group 1 *Theraps* contains those species that prefer a slow to moderate current or flow of water. In the lower to middle reaches of the river valley the water has a higher level of oxygen and a slightly lower temperature. The food available is different from the previous group, as there is much less phytoplankton and zooplankton in the rivers than in the lakes. *Aufwuchs* and other types of filamentous algae replace the lower forms of algae found in the lakes. The live foods available in the lower and middle river valleys have much less fish content, but is more generous in insects and invertebrates.

These slightly elongated species are not as large or as deep-bodied as their more aggressive lacustrine cousins. Their mouths are smaller and located near the bottom of their heads for collecting food from the river bottom. They prefer to lay their eggs on a rock substrate, rather than hardened earth, and their spawns are not as great in number.

GROUP 1A

C. bifasciatum   C. heterospilus
C. maculicauda   C. melanurum
C. synspilum   C. zonatum

GROUP 1B

C. argentea   C. breidohri
C. fenestratum   C. "Conkel"
C. guttulatum   C. hartwegi
C. micropthalmus   C. nicaraguense
C. regani   C. sp. "Brindisi"

C. (T.) bifasciatum (STEINDACHNER, 1864)

COMMON NAME: Red speck cichlid

DISTRIBUTION: Southern Mexico and northern Guatemala on the Atlantic slope. Restricted to the Rio Usumacinta basin.

HABITAT: pH: 7.0-7.8, GH: 3-14, KH: 3-8, T: 26-30° C. This large *Theraps* species prefers the calm waters of the lakes and lagoons and the lower river valleys with slow to moderate flowing current. Their diet consists mainly of bottom detritus and vegetable matter, but they will eat most all forms of plant life, from algal filaments to terrestrial plants.

Cichlasoma (Theraps) bifasciatum. Rio Tulija, Mexico.

DISTINCTIVE CHARACTERISTICS: SL: 300 mm. *C. (T.) bifasciatum* has two broad black horizontal stripes on the side. The lower stripe is wide and solid, extending from the pectoral fin base to the base of the caudal fin, with a slight upward curve. The upper stripe is somewhat blotched, joins with the lower stripe at its origin, and extends toward the end of the dorsal fin base. Their juvenile color is the typical drab gray which most all the *Theraps* exhibit. The adult male's body coloration is yellow-orange, and this color extends into the base of its fins, where it blends into an iridescent turquoise green. The head, shoulder, and breast area are bright red, and bright red specks cover the orange face. The fins are spangled in maroon stripes and spots and tipped or edged in red. The females are similarly colored, but lack the intensity. They lack the heavy red speckling and maroon spangling.

C. (T.) breidohri (WERNER & STAWIKOWSKI, 1987)

DISTRIBUTION: This species is known only on the Pacific slope of Mexico in Presa de la Angostura tributaries, from Villa Flores, Chiapas, Mexico, to Rio Lagartero, Huehuetenango, in extreme western Guatemala.

HABITAT: pH: 7.5-8.0, GH: 10-20, KH: 12-18, T: 24-30° C. This species exists in the lower to middle river valleys as well as the lake, Presa de Angostura. It can be found over rocks, sand, silt, or mud. It inhabits most all the habitats available within its small territorial boundaries.

DISTINCTIVE CHARACTERISTICS: SL: 250 mm.

C. (T.) fenestratum (GUENTHER, 1860)

SYNONYMS: *Heros parma* Guenther, 1862, *Cichlasoma sexfasciatus* Regan, 1905, *Cichlasoma gadovii* Regan, 1905

DISTRIBUTION: Mexico, on the Atlantic slope. Restricted to the Rio Coatzacoalcos basin, in the states of Oaxaca and Veracruz.

HABITAT: pH: 7.0-7.5, GH: 6-8, KH: 8-10, T: 23-30° C. This species is generally lacustrine, proliferating in the mesotrophic waters with a mud or sand bottom. Some populations have migrated into the lower river valleys with slow to moderate flowing waters and have adapted with a more elongated body to facilitate thrust and movement in the flowing water. They grow to about 10-12 inches under lacustrine conditions and slightly smaller under riverine conditions. They eat mainly plant and vegetable matter, but supplement their diet with invertebrates and aquatic insects. Even in the lakes, these fishes prefer the submersed large boulders and trees. They congregate in heavy numbers near these reefs.

DISTINCTIVE CHARACTERISTICS: SL: 250 mm. The males have a beautiful dark turquoise coloration that fades into a sulfur yellow color in the upper back and into the dorsal fin. A wide black band extends through the body just below the lateral line. During sexual activities five vertical bars appear, and the eyes glow a bright sulfur color.

Breeding pairs are not particular about where they nest. They will lay their eggs on wood, rock, or in the sand. Their territories in nature are small for cichlids of their size, about 4 cubic feet, but they are very aggressive.

Cichlasoma (Theraps) breidohri. **Prese de la Angostura, Mexico.**

Cichlasoma (Theraps) fenestratum. Lago Catemaco, Mexico.

They will chase intruders away for 20-30 feet before returning to the brood site.

C. (T.) sp. "Conkel"

COMMON NAME: Red fenestratum

DISTRIBUTION: Mexico, on the Atlantic slope. This species is endemic to Lago de Catemaco, Veracruz.

HABITAT: pH: 7.4, GH: 8, KH: 10, T: 23-29° C. This species is confined to the rocky areas of this volcanic lake. They graze upon the *aufwuchs* and crustaceans.

DISTINCTIVE CHARACTERISTICS: SL: 230 mm. As a juvenile this species is the typical gray coloration with a solid wide bar along the lateral line. As adults, they develop extreme differences in coloration, depending on the locality in the lake. Some populations are pink, some red, and some white, while many are piebald, displaying all three colors. Some are spotted in black, and many have sulphur colored fins.

C. fenestratum is found sympatric with this species. The differences are as follows: this species is more elongated and not as deep bodied; its mouth protrudes forward and

Cichlasoma (Theraps) sp. "Conkel". Lago Catemaco, Mexico.

ALL FOUR:
Cichlasoma (Theraps) sp."Conkel." Lago Catemaco, Mexico.

the head is not as convex; as juvenile, the body has a purple cast and a thinner lateral stripe, which is sometimes broken. At about three inches this unidentified species begins to change colors, much like *C. citrinellum* and *C. labiatum.*

C. (T.) *hartwegi* TAYLOR & MILLER, 1980

DISTRIBUTION: Mexico, on the Pacific slope. Restricted to the Rio Grande de Chiapa and its tributaries, including the Presa de Angostura.

HABITAT: pH: 7.5-8.0, GH: 10-20, KH: 12-18, T: 24-30° C. This species exists in the lower river to middle river valleys, as well as the lake, Presa de Angostura. It can be found over rocks, sand, silt, or mud. It inhabits most all the habitats available within its small territorial boundaries. It has a widely varied diet, consuming bottom detritus, aquatic plants, and invertebrates.

DISTINCTIVE CHARACTERISTICS: SL: 250 mm. All sizes have a thin, disrupted longitudinal stripe that extends from the center of the pectoral fin to a spot centered at the base of the caudal fin. A large longitudinal blotch exists in the middle of the body, just above this bold stripe. Its head profile is convex, with a slightly inferior mouth. The upper lip is thicker and protrudes in front of the lower lip. Its basic body coloration is a metallic silver to blue-silver, with series of small vertical wavy lines and spots arranged on each scale and throughout all the fins on its body. A thin interorbital stripe across its head links its two eyes.

C. (T.) heterospilus HUBBS, 1936

DISTRIBUTION: Southern Mexico and northern Guatemala, on the Atlantic slope. Restricted to the lower and middle reaches of the Rio Usumacinta basin.

HABITAT: pH: 7.0-7.5, GH: 3-5, KH: 3-8, T: 26-30° C.

DISTINCTIVE CHARACTERISTICS: SL: 240 mm.

C. (T.) maculicauda REGAN, 1905

SYNONYMS: *Cichlasoma globosum* Miller, 1907, *C. manana* Miller, 1907, *C. nigritum* Meek, 1907, *Vieja panamensis* Fernandez-Yepez, 1969

COMMON NAME: Black-belt cichlid

DISTRIBUTION: From northern Belize in the Belize River to the Rio Chagres in Panama,

Cichlasoma (Theraps) heterospilus. Photo by Heijns.

Cichlasoma (Theraps) hartwegi. Rio Grande de Chiapas, Mexico.

Cichlasoma (Theraps) maculicauda. Rio Sarstoon, Belize/ Guatemala.

on the Atlantic slope. It also is found in Lago de Nicaragua.

HABITAT: pH: 6.8-8.2, GH: 3-8, KH: 3-6, T: 26-32° C. This is an estuarine species and has the widest distribution of all Central American cichlids due to its ability to tolerate brackish and marine conditions. It is decidedly lacustrine, but will migrate into the extreme lower sections of the lower river valleys where the current is slow. I have collected specimens along the mangrove shorelines of Corn Island, some 30 miles from the mainland of Nicaragua in the Atlantic Ocean. Their main food source is the bottom detritus, consisting of vegetable matter, both aquatic and terrestrial plants, seeds, and fruits. They prefer muddy or sandy bottoms and live among the submersed trees and logs for protection. In riverine conditions, they prefer the shady bank edges, where they forage upon the aquatic and terrestrial plants.

DISTINCTIVE CHARACTERISTICS: SL: 300 mm. Individuals of this species always have a large black spot on the caudal peduncle. Juvenile patterns and fright patterns are made up of 5 vertical bars, with the middle bar widened more than the others. As adults, they still display this middle bar, thus giving them the common name black-belt cichlid. Their body coloration as juveniles is the typical drab gray, the color of most all *Theraps* species, with transparent fins. As adults, they transform into a green to blue-green dress with an overcast of copper shades. Varying shades of red exist on the face and head of mature specimens, as well as on the fins. Small black freckles of irregular size and shape predominately appear on the males.

C. maculicauda is an open water spawner and has about 600-1000 fry per spawn. Both parents participate in the care of the young, and the males stay with the family for 4-5 weeks. This is a long time for American cichlids.

C. (T.) melanurum (GUENTHER, 1862)

DISTRIBUTION: Guatemala, on the Atlantic slope. Endemic to the basin of Lago Peten.

HABITAT: pH: 7.4-7.6, GH: 13-15, KH: 5-6, T: 26-30° C. This species is typically lacustrine, but prefers a sandy bottom as compared to a rocky bottom. It is not as large or deep bodied as its closest relative, *C. synspilum*. Its food source is *aufwuchs*, bottom detritus of vegetable matter, and small crustaceans.

DISTINCTIVE CHARACTERISTICS: SL: 190 mm. The gray juvenile

**Cichlasoma (Theraps) melanurum.
Lago Peten Itza , Guatemala.**

coloration is enhanced by a purple hue, and a solid black lateral bar extends from the tip of the caudal peduncle forward to about the middle of the body. An irregular dark splotching occurs at the base of the dorsal fin, forming a second horizontal bar. Both adult males and females exhibit a bright orange coloration, with blue extending into the anal and caudal fins. During courtship the eyes become bright gold, and the lower body turns a deep blue-black.

C. melanurum is rather difficult to spawn and reproduces in small quantities of 300-500 fry per spawn. This species is very popular because of its bright colors, its relatively small size, and its behavior, which is much less aggressive with conspecifics than most of the other *Theraps* species.

C. (T.) nicaraguense (GUENTHER, 1864)

SYNONYMS: *Cichlasoma balteatum* Gill & Bransford, 1877, *Cichlasoma spilotum* Meek, 1912

NATIVE NAME: *moga*

DISTRIBUTION: Nicaragua and Costa Rica, on the Pacific and Atlantic slopes. This species is the third most frequently found cichlid in the Great Lakes of Nicaragua. It also exists in the Rio San Juan system and extends south to the Rio Matina on the Atlantic

Cichlasoma (Theraps) nicaraguense. Female. Rio Cuba, Costa Rica.

slope of Costa Rica.

HABITAT: pH: 7.5-8.8, GH: 2-5, KH: 2.0-4.5, T: 26-32° C. *C. nicaraguense* are found in the lakes and the slow moving rivers with moderate currents. The juveniles feed on aquatic insects, while the adults feed on the bottom detritus and sand, seeds, and leaves.

DISTINCTIVE CHARACTERISTICS: SL: 200 mm. This elongated *Theraps* species is distinguished by the black spot on the mid-lateral section of the body. This large spot is stretched vertically and sometimes disappears with age. The head profile is extremely curved and its small mouth is placed on the inferior section or bottom of the head. Their body coloration is gold to copper with a blue-green to gray head, depending on the locality. The males have scales that are bordered in a dark opaque color, forming a reticulated pattern. With the onset of sexual maturity, this species becomes clearly dimorphic, with the intensification of color from gold to orange. Males grow much larger than the females and exhibit a nuchal hump. They also have many dark spots in their fins, and their dorsal fin is edged in red. The females are solid orange with no reticulated pattern, and their fins are void of dark spotting.

This species lays its eggs in depressions in the sandy bottom rather than adhering them to submersed rocks or tree trunks and limbs, as do most cichlids. Its spawns number about 200-400, with the females practicing a rather

unique communal care during the post spawning period. Termed **creching**, a group of three or four females stand guard over their combined spawns, encircling the expanded group, and do not allow any intruders or predators into the rearing arena. This is unusual for cichlid behavior, but I have noted it in *C. altifrons* and *C. rostratum*.

Cichlasoma (Theraps) regani. Rio Jaltepec, Mexico.

bars, barely visible, which disappear with age. Two to three dark spots or blotches are visible just above the lateral line, with the largest blotch being the anterior one. A black blotch occurs on most of the caudal peduncle. Adult males have a light green body coloration that is freckled with numerous small red specks, 2-4 per scale. This red freckling extends to the operculum

C. (T.) regani MILLER, 1974

NATIVE NAME: *mojarra pinto*

DISTRIBUTION: Southern Mexico, Atlantic slope. Endemic to the upper reaches of the Rio Coatzocoalcos basin in Mexico, in the rivers Rio Almoloya, Rio Jaltepec, Rio Malotengo, and Rio Sarabia in the state of Oaxaca.

HABITAT: pH: 7.2-7.5, GH: 2-5, KH: 4-6, T: 24-26° C. Their diet is mostly vegetable bottom debris and *aufwuchs*.

DISTINCTIVE CHARACTERISTICS: SL: 230 mm. The juveniles have up to 7 thin vertical

and the lower facial area. Two parallel and saddle-shaped bars extend between the eyes, forming a mask. All of the fins are heavily speckled. The females are freckled as well, but in a blue, rather than a red, color.

C. (T.) synspilum HUBBS, 1935

SYNONYMS: *Cichlasoma hicklingi* Fowler, 1956

COMMON NAME: Pastel cichlid

DISTRIBUTION: Mexico, Guatemala and Belize, on the Atlantic slope. Restricted in the Rio Usumacinta basin,

Belize River, Belize.

Cichlasoma (Theraps) synspilum.

Progresso Lagoon, Belize.

Rio Tulija, Mexico.

from the Rio Tulija in Tabasco, Mexico, to Clearwater Creek in southern Belize.

HABITAT: pH: 7.0-7.8, GH: 3-9, KH: 3-8, T: 26-30° C. Typical lacustrine and lower river valley species with slight tolerance of brackish water or the estuarine environment.

DISTINCTIVE CHARACTERISTICS: SL: 300 mm. Juveniles have typical gray coloration, with a green hue and a broken lateral stripe that extends from the caudal peduncle forward to about one third the length of the body. Mature males produce nuchal humps,

the largest of all the American species. As adults, this species exhibits extreme variation in color patterns, but the most common form displays a bright orange body coloration with turquoise hues on the undersides and throughout the fins. A large percentage of the scales (30-50%) are black or are bordered in black. In some varieties the other scales are alternatively black and orange, forming a symmetrical, reticulated pattern. The fins also are spotted and streaked in black. With a bright red face and head, this species is one of the most popular in aquarium circles.

In central and southern Belize, an overall orange color variant exists, similar to *C. melanurum*. The throat and lower facial area have a light pink color. In Chiapas, Mexico, a turquoise color variant is found in the highlands of the Rio Tulija. However, in the more turbid waters of the lowland stretches of this same river, the orange variant is the only form found. This color differentiation probably is attributed to the dietary differences of the two races.

C. (T.) zonatum MEEK, 1905

NATIVE NAME: *mojarra prieta*

DISTRIBUTION: Mexico, on the Atlantic slope. Restricted to the Rio Coatzacoalcos basin in the Rio Jaltepec to the Rio Grijalva basin in the Rio Teapa.

HABITAT: pH: 7.2-7.5, GH: 2-5, KH: 4-6, T: 23-30° C.

DISTINCTIVE CHARACTERISTICS: SL: 250 mm. Juveniles exhibit the typical gray coloration, with a blue cast or hue. A broad solid bar extends from the caudal peduncle to the gill cover. An irregular, black shaded area exists in the middle of the body, just above this lateral bar. As adults, both sexes of this species turn a bright blue, with a sulfur overcast in the upper body and the dorsal and anal fins. The area below the lateral line turns a solid white, which extends forward to the face

Cichlasoma (Theraps) zonatum. Rio Joltepec, Mexico.

and lips. A green patch extends from the eye and into the lips. Small black freckling is found in both the dorsal and anal fins, but the anal, pectoral, and pelvic fins are almost transparent.

THERAPS: GROUP 2

The second group of *Theraps* species are riverine in nature and occupy the upper reaches of cichlid habitat. They are smaller, thinner, and have elongated bodies to accommodate their maneuverability in the faster flowing waters of the highlands. These thin bodied cichlids require highly oxygenated, fast flowing water in which to live. The waters in these upper river sections are high in pH (7.5-9.0) and mineral content, and are much cooler, 22-26° C. in temperature. In this mountain brook environment, they dart around the bottom, moving from one rock or wood shelter to another. They eat much like the mbuna from Lake Malawi, in that they graze on the *aufwuchs* that cover the submersed trees, rocks, and bottom. *Aufwuchs* is rich in protein, containing algae and various invertebrates. Nicknamed by the native Indians *corrientera* and by the Hispanic people *currentos*, both meaning current-dweller, these cichlids are for the

Cichlasoma (Theraps) coeruleus. Photo by R. Stawikowski.

Cichlasoma (Theraps) coeruleus. Photo by R. Stawikowski.

advanced hobbyists who are capable of maintaining rare and delicate fishes. They also are the preferred Neotropicals by the African cichlid devotees, as they have similar traits of smaller size, delicate shape, bright colors, and similar feeding habits. They are like the Malawi mbuna and Tanganyikan *Tropheus* in that they cannot tolerate high concentrations of nitrates, nitrites, or ammonia. Always keep a good mechanical filter and pump with these fishes, as they will prosper and breed only with clean, flowing water.

C. (T.) coeruleus (STAWIKOWSKI & WERNER, 1987)

SYNONYMS: *Cichlasoma omonti; Cichlasoma belone*
NATIVE NAME: Its Chol Indian name is *barsin*, meaning "vertical bars."
DISTRIBUTION: The Atlantic slope of Mexico. The upper reaches of the Usumacinta system in the Rio Mi-sol-Ha and Rio Shamula, the montane affluents of the Rio Tulija.
HABITAT: pH: 7.5-8.0, GH: 13-18, KH: 14-18, T: 23-25° C.
DISTINCTIVE CHARACTERISTICS: SL: 120 mm. This species is the smallest, and perhaps most peaceful, of the riverine *Theraps*. The body is powder blue with a light green belly, throat, and gill cover. The

pattern of bars is unique in that a "half moon" shaped horizontal band rests atop the second and third vertical bar, forming a "saddle." Two cross bands connect the eyes and a third wraps around the forehead area. The fins are violet to red and overlaid with bright blue spangling. This current-loving fish prefers the montane habitat and thrives upon the *aufwuchs* and the encrusted invertebrates (mostly small crustaceans). This species might be termed a dwarf cichlid, as not one that was collected was larger than 120 mm.

C. (T.) gibbiceps STEINDACHNER, 1864

SYNONYM: *Cichlasoma teapae* Evermann & Goldsborough, 1902
NATIVE NAME: *Roquera* (meaning "rock dweller")
DISTRIBUTION: Atlantic slope of southern Mexico. Restricted to the reaches of the Rio Grijalva basin.
HABITAT: pH: 7.2-7.5, GH: 2-8, KH: 4-7, T: 23-26° C.
DISTINCTIVE CHARACTERISTICS: SL: 230 mm. This cylindrical species has a pattern of six black vertical bars that extend from the dorsal fin to just

Cichlasoma (Theraps) gibbiceps.

Cichlasoma (Theraps) godmanni. Rio Dulce, Guatemala.

below the mid-lateral line. A large black spot in the middle of the mid-lateral line on the third bar is easily identifiable. Two cross bands extend between the eyes. The metallic green body and fins are covered with hundreds of small red spots or freckles, giving it a most unusual and interesting pattern. When the fish is young the spots are maroon in color, but as it matures they turn a bright red.

C. (T.) godmanni (GUENTHER, 1862)

DISTRIBUTION: Guatemala, on the Atlantic slope. Restricted to the upper reaches of the Lake Izabal basin, in the Rio Polochic, Rio Cahabon, and Rio Dulce and their tributaries.

HABITAT: pH: 7.3-8.0, GH: 4-9, KH: 3-6, T: 26-30° C. This species is strictly riverine, existing in the affluents of Lago de Izabal and its effluent, the Rio Dulce. Here they prefer the fast flowing, cooler waters (10-15° F cooler) of the mountain runoffs. They spawn on the sides of larger rocks and in caves that face away from the swift current. Their eggs are yellow-green in color, rather hard shelled, and extremely adhesive. Rarely do they congregate together as adults. Instead, they prefer to exist alone or in pairs.

DISTINCTIVE CHARACTERISTICS: SL: 200 mm. This species is easily identifiable by its bold check-mark pattern that exists throughout its life. A broad black band runs from a spot at the top of the gill cover, angling slightly downward to the lower flanks, and then stops. A vertical bar connects at that point and runs upward at about a 10-15 degree angle toward the tail end of the dorsal fin. A second, smaller, vertical bar runs behind and parallel to this bold bar. Light black freckling is evident in the face and nape. Color variations do occur, with a turquoise form existing in the Rio Polochic and its tributaries, and a red form in two effluents of the Rio Dulce. The red form has brilliant turquoise spangling on its body and fins and is very rare in nature.

Even though this species hails from cool waters, it is very susceptible to ich.

C. (T.) guttulatum (GUENTHER, 1864)

DISTRIBUTION: Southern Mexico to northern Guatemala, on the Pacific slope. Rio Chicapa and Rio Tehuantepec in Oaxaca, Mexico, to Lago Amatitlan, Guatemala.

HABITAT: pH: 7.5-8.0, GH: 10-20, KH: 12-18, T: 24-30° C. A riverine species that lives in lower, middle, and upper river valleys across a broad scope of substrate, from mud to sand to rocks.

DISTINCTIVE CHARACTERISTICS: SL: 190 mm. This elongated *Theraps* has a convex head and inferior mouth. A broad longitudinal line lies just below the lateral line, from behind the pectoral fins to the caudal. This *Theraps* has a metallic turquoise coloration on the body and throughout the fins. A bright blue bar runs from the eye to the upper part of the operculum and

across the forehead, giving it an unusual colored mask. Dark maroon spots lie in symmetrical horizontal rows, giving this species a reticulated pattern. These spots continue into the dorsal, caudal, and anal fins. The pectorals and pelvics are transparent.

C. (T.) intermedium (GUENTHER, 1862)

SYNONYMS: *Heros angulifer* Guenther, 1862, *Heros rectangulare* Steindachner, 1864

DISTRIBUTION: Southern Mexico, northern Guatemala, and Belize, on the Atlantic slope. The upper reaches of the Rio Usumacinta basin in the Rio Tulija and Rio Mi-sol-Ha in Chiapas, Mexico, to the Rio Sarstoon, the border of Belize and Guatemala.

HABITAT: pH: 7.5-8.0, GH: 13-18, KH: 14-18, T: 26-30° C. This species prefers the moderate flowing waters of the middle and upper river valleys. They move up and down the river in small sized groups, usually 10-30, staying relatively close to the larger rock formations.

DISTINCTIVE CHARACTERISTICS: SL: 200 mm. Juveniles and adults are identified easily by the angled check-mark pattern on their sides. A broad black band runs from a spot at the top of the gill cover, angling slightly downward to the lower flanks, and then stops. A vertical bar connects at about a 10-15 degree angle toward the tail end of the dorsal fin. A light black freckling is evident on the face, nape, and back. Color variations occur and may be considered sibling species as suggested by Guenther and

Cichlasoma (Theraps) guttulatum. Rio Chicapa, Mexico.

Cichlasoma (Theraps) intermedium. Rio Moho, Guatemala.

Regan. To the north, in southern Mexico, the colors are the usual olive-gray, with a green to blue body overcast and fin spangling. The black freckling is prominent, and some of the scales on the nape and back are bordered in black. The specimens in Belize are rather drab, with olive-gray bodies and light freckling, while their cousins to the west in Guatemala, including Lake Peten, have bright pink to orange sides and fins.

For spawning, this species must be mature and rather large, considering their medium size. At about three years of age, a female of about 7-8 inches will produce about 300-400 eggs. Their contrasting black and white breeding pattern is a sure sign of reproductive maturity. These fishes are best kept in harems of one male to three or four females, but become rather nasty to all intruders during courtship. It is important to remove the subdominant females to another aquarium during this time and reintroduce the spawning male with them after the eggs hatch.

C. (T.) irregulare (GUENTHER, 1862)

DISTRIBUTION: Southern Mexico and Guatemala, on the Atlantic slope. The upper reaches of the Grijalva basin in the Rio Santa Cruz, to the upper reaches of the Lago Izabal basin in the Rio Polochic and the Rio Dulce and their tributaries.

HABITAT: pH: 7.3-8.0, GH: 4-9, KH: 3-6, T: 26-30° C. Exists only in the fast moving waters of the upper river valleys. Enjoys the rocky habitat and prefers the sections of the rivers where the current is strong with white water. This species thrives on the *aufwuchs*, from which it gets its nourishment from the algae and invertebrates.

DISTINCTIVE CHARACTERISTICS: SL: 190 mm. This slender bodied *corrientera* behaves much like a goby or a darter, as it propels itself through the water with the greatest of speed and ease. It then settles down on the river bottom in a resting position, with its pectoral fins spread out, and forages on the *aufwuchs* or the invertebrates and small crustaceans between the rocks and stones. Its body is cigar shaped, with a prominent, compressed snout. Its mouth is inferior, and the upper lip protrudes past its lower lip. A longitudinal, broken, black stripe extends from just above the pectoral fin into the caudal fin itself. Three prominent black blotches appear on the back at the base of the dorsal fin. Juveniles, as well as adults, are a butterscotch color. Adults have a rosy red colora-

Cichlasoma (Theraps) irregulare. Rio Cienega, Guatemala.

tion on the face, throat, and midsection, above the lateral line. A patch of robin's egg blue color is located just behind the pectorals and just above the pelvic fins, and extends to the caudal below the lateral stripe. Large spangles and irregular stripes of this metallic blue cover the operculum and encircle both eyes. During courtship and spawning a large section of black covers the breast and belly, and a black mask appears.

This species is most definitely aggressive to its conspecifics. The males prefer the largest of the females to breed with, and all others should be removed at the onset of pair formation and courtship. *C. irregulare* is prone to ich, so keep the temperature at a steady 25-26° C.

C. (T.) lentiginosum (STEINDACHNER, 1864)

SYNONYM: *Theraps rheophilus* Seegers & Staeck, 1985

DISTRIBUTION: Southern Mexico and northern Guatemala, on the Atlantic slope. Upper tributaries of the Rio Usumacinta system in Mexico and Guatemala.

HABITAT: pH: 7.5-8.0, GH: 13-18, KH: 14-18, T: 26-30° C. The usual preference of cooler, fast flowing water of high mineral content. The usual diet of *aufwuchs* and invertebrates.

DISTINCTIVE CHARACTERISTICS: SL: 200 mm. This elongated species is similar to the above species, but a bit taller and with a more convex head. As a juvenile, six spots appear along the mid-lateral line. These spots are the bottom base for faint vertical bars that appear on the juveniles

and on frightened adults. Heavy black freckling covers the upper half of the body, the dorsal fin, the gill cover, and the area beneath the eye. As adults, the body turns a light blue color, with deeper blue in the fins. The gill cover and face turn a deep green. A light maroon color streaks the pelvic, anal, and caudal fins only. Two interorbital stripes link the eyes. A nuchal hump is prominent in older males. Females have only a slight amount of freckling, and a deep blue blotch appears in the middle of the dorsal fin when maturity approaches. A washed out white color is its breeding dress for both sexes.

C. (T.) micropthalmus (GUENTHER, 1862)

SYNONYMS: *Heros oblongus* Guenther, 1866, *Cichlasoma guentheri* Pellegrin, 1904, *Cichlasoma milleri* Meek, 1907, *Cichlasoma caeruleogula* Fowler, 1935

DISTRIBUTION: Southern Guatemala and northern Honduras, on the Atlantic

Cichlasoma (Theraps) lentiginosum. Rio El Subin, Mexico.

slope. Endemic to the Rio Motagua basin, including Lago Amatitlan.

HABITAT: pH: 7.0-7.3, GH: 6-10, KH: 4-10, T: 26-28° C.

DISTINCTIVE CHARACTERISTICS: SL: 250 mm.

Cichlasoma (Theraps) micropthalmus. Photo by Hejins.

C. (T.) nebuliferum (GUENTHER, 1860)

NATIVE NAME: *playero* (meaning beach or gravel dweller)

DISTRIBUTION: Rio San Juan drainage south of the Rio Papaloapan.

HABITAT: pH: 7.0-7.5, GH: 6-8, KH: 8-10, T: 23-30° C.

DISTINCTIVE CHARACTERISTICS: SL: 200 mm. This elongated species of the subgenus *Theraps* displays a black lateral band extending from the eye to the ocellated spot in the caudal peduncle. The body is metallic green, which varies from light to dark, depending upon its mood. Bright yellow scales are scattered intermittently throughout the upper body and dorsal fin. A rose pink color is displayed beneath the lateral band, from the lips to the caudal fin.

C. (T.) panamense (MEEK & HILDEBRAND, 1913)

NATIVE NAME: *chogorro*

DISTRIBUTION: Panama, on the Pacific slope. Rio Chagres, Rio Tuira, and Rio Bayano.

HABITAT: pH: 7.0-7.9, GH: 3-4, KH: 3-4, T: 26-30° C. This

Cichlasoma (Theraps) nebuliferum.

species prefers moderately flowing water over a substrate of small stones, sand, rocks, and leaf debris from the forest canopy. A layer of flocculent green organic material, *aufwuchs*, provides abundant nourishment.

Cichlasoma (Theraps) panamense. Rio Chagres, Panama.

DISTINCTIVE CHARACTERISTICS: SL: 130 mm. A series of seven to nine irregularly shaped black blotches are present along the side of the body, just below the mid-line, and three to four appear on the anterior portion of the lateral line. The typical base color is olive gray, with the head and dorsal surface a green color. On the sides, around the pectoral fin, is a large reddish patch that extends onto the throat. In some populations, this red patch is quite extensive. Females possess a black blotch in the center of the dorsal fin.

This is the smallest of the *Theraps* species and one of the more difficult to spawn in captivity. The fry are extremely small and must be provided with infusoria for development.

C. (T.) sieboldii (KNER & STEINDACHNER, 1863)

SYNONYMS: *Heros montezuma* Heckel, 1840, *Cichlasoma*
frontale Meek, 1909, *C. frontosa* Meek, 1909, *C. punctatum* Meek, 1909, *Theraps terrabae* Jordan & Evermann, 1927

DISTRIBUTION: Costa Rica and Panama, on the Pacific slope. From the Rio Jesus Maria in Costa Rica to the Rio Santa Marta in Panama.

HABITAT: pH: 7.5-8.2, GH: 2-6, KH: 2-6, T: 24-29° C. This elongated riverine species requires moderate to fast flowing waters and is tolerant of the cooler mountain water temperatures (24-29 °C.). Juveniles feed primarily upon aquatic insects, while adults feed on vegetable matter, such as *aufwuchs* and algae.

DISTINCTIVE CHARACTERISTICS: SL: 250 mm. This species has two parallel horizontal bars made up of broken oval splotches. The mid-lateral line is the most obvious of the two, with six spots forming a bar that extends onto the gill cover. A mask is formed on its convex forehead by two paral-
lel bands linking the eyes. The mouth is situated slightly on the inferior side of its convex head. Juveniles are gray, while adults have a body coloration of olive-green above the mid-lateral line and white below the line extending throughout the abdomen. Males are lightly freckled with small black dots or spots above the mid-lateral line and maroon red dots or spots below it in marked serial rows. The fins of the male are freckled also with black and maroon spotting.

This fish is rather difficult to spawn and thus rare among hobbyist circles. This species spawns during the hot and dry part of the year, with fry numbers ranging from 200-300. The contrasting black and white coloration of the breeding dress is very attractive.

C. (T.) tuba MEEK, 1912

SYNONYM: *Tomocichla underwoodi* Regan, 1908

DISTRIBUTION: Costa Rica and Panama, Atlantic slope. Ranges from the Rio Escondido in Nicaragua to the Rio Cricamola in northern Panama.

HABITAT: pH: 7.2-7.6, GH: 3-8, KH: 3-8, T: 24-29° C. This species is found in the moderate to fast flowing waters of rivers with high velocity. This species's diet consists of aquatic plants, algae, and fruits. The juveniles are partial to aquatic insects.

DISTINCTIVE CHARACTERISTICS: SL: 240 mm. The body is elongated, the head is convex, and the mouth is inferior. The lateral band is a series of 9 vertical bars from the operculum to the caudal. The area above the lateral stripe is gray

Cichlasoma (Theraps) sieboldii. Rio Esquinas, Costa Rica.

Cichlasoma (Theraps) tuba **Rio San Juan, Nicaragua/Costa Rica.**

body, above the mid-lateral line. Three to four specks are arranged irregularly on each body scale, giving it a most unusual and interesting pattern. Only the pectoral fins are void of this "peppered" appearance. Two stripes extend between the eyes and a charcoal black "mask" or "crown" is visible on the forehead. The females attain a smaller size and do not have as much black freckling. The males develop an exaggerated nuchal hump during periods of sexual behavior.

C. (T.) sp. "Brindisi"

DISTRIBUTION: On the Atlantic slope of southeastern Mexico. It is endemic to the Rio Tulija and its affluents south of Salto de Agua, Chiapas.

HABITAT: pH: 7.0-7.8, GH: 3-9, KH: 3-8, T: 26-30° C.

DISTINCTIVE CHARACTERISTICS: SL: 200 mm. The juveniles have five barely visible, broad vertical bars that disappear with age. Two irregular black blotches, each about 20 mm in diameter, are visible. One is located on the caudal peduncle, the other in the upper nape region just below the

to black and the area below the lateral stripe is white. The area between the vertical bars on the mid-lateral stripe is a light purple. This purple area extends from the top of the operculum across the forehead. Two black horizontal lines link the eyes, similar to those in *C. sieboldii*. The dorsal fin is black to gray, the pelvic fins are white, and the anal and caudal fins are transparent. Older males sometimes possess a nuchal hump. The juveniles have 9 black vertical bars on a gray body or base color.

C. (T.) argentea (ALLGAYER, 1991)

DISTRIBUTION: On the Atlantic slope of southeastern Mexico. This form is restricted to the southeastern sections of the Rio Usumacinta basin.

HABITAT: pH: 7.0-7.8, GH: 3-14, KH: 3-8, T: 26-30° C.

DISTINCTIVE CHARACTERISTICS: SL: 300 mm. This species has the typical body shape of species of the *Theraps* group. It is silver in color, appearing pearlescent in certain light. Two large blotches are displayed, one on the caudal peduncle and one slightly forward of the middle of the

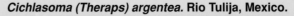

Cichlasoma (Theraps) argentea. **Rio Tulija, Mexico.**

Cichlasoma(Theraps) sp. "Brindisi."

anterior spines of the dorsal fin. The body and fins are lavender, with each scale freckled with black spots, about 5-6 per scale. These spots are found in the dorsal and caudal fins, as well as on the bases of the opaque anal, pectoral, and pelvic fins. This black freckling also is seen on the gill cover and lower facial area. Two parallel stripes extend between the eyes, forming a mask, with the upper stripe continuing on an area just below the eye. A triangular shaped "mask" or "crown" occasionally is visible on the forehead. The dorsal fin is very high when compared with other members of the subgenus *Theraps*, much more like that of species in the *aureum*-complex of the subgenus *Thorichthys*.

C. (T.) sp. "Tulija"

DISTRIBUTION: Atlantic slope of southeastern Mexico. This form is limited to the middle sections of the western Rio Grijalva system and southeastern sections of the Usumacinta basin.

HABITAT: pH; 7.0-7.8, GH: 3-9, KH: 3-8, T: 26-30° C.

DISTINCTIVE CHARACTERISTICS: SL: 200 mm. This rare form has an elongated body that is silvery in coloration, but appears pearlescent in certain lighting. Different hues of blue, green, and violet can be seen when light is reflected from different angles. Two large black blotches are displayed, one in the middle of the longitudinal line and the other on the caudal peduncle.

The body is freckled with numerous black spots in an asymmetrical pattern. These spots extend into the dorsal and caudal fins. The anal and pelvic fins are streaked with lavender. Six barely visible black vertical bars extend from the dorsal fin to the mid-lateral line.

C. (T.) sp. "Chomba"

DISTRIBUTION: On the Atlantic slope of southeastern Mexico. Collected in the Rio San Joaquin, the headwaters of the Rio Chumpan, in the state of Campeche.

Cichlasoma (Theraps) sp. "Chomba."

HABITAT: pH: 7.2-7.8, GH: 6-14, KH: 4-7, T: 26-30° C.

DISTINCTIVE CHARACTERISTICS: SL: 250 mm. In this middle and lower riverine species the

Cichlasoma (Theraps) sp. "Tulija."

Cichlasoma (Nandopsis) sp.
"Darsti."

head and shoulder regions are olive green while the rest of the body is charcoal black. Large, irregular spots of turquoise adorn the body in a non-uniform pattern, but symmetrically arranged in the dorsal, anal, and caudal fins into a "checkerboard" or "lace-like" pattern. Only the fore-head and areas around the nostrils lack the spotting. This interesting feature gives the fish a touch of elegance.

C. (N.) sp. "Darsti"

DISTRIBUTION: On the Atlantic slope of southeastern Nicaragua. Collected in the Rio Escondido near Bluefields.

HABITAT: pH: 7.0-7.8, GH: 3-14, KH: 3-8, T: 26-30° C.

DISTINCTIVE CHARACTERISTICS: SL: 200 mm. This species of the subgenus *Nandopsis* is riverine, preferring the deeper holes of the middle and lower valleys where the water is cooler and more oxygenated than that of the lakes. The primary diet consists of aquatic and terrestrial insects, small crustaceans, and small fishes. Its body is beige in color, with six dark green vertical bars. The anterior bar is split at the mid-line and forms a saddle over the gill cover. Two broad stripes cross the forehead, one between the eyes and another just above the eyes connecting the spots

on the gill cover. Two large blotches occur on the mid-line, one in the middle of the body and one on the caudal peduncle. The maroon dorsal, anal, and pelvic fins are streaked in blue, while the caudal fin is spangled with blue near an ocellated spot on the caudal peduncle.

C. (T.) sp. "Nicapa"

DISTRIBUTION: Found on the Pacific slope of southern Mexico in the Rio Colorado.

HABITAT: pH: 7.5-8.0, GH: 10-20, KH: 12-18, T: 24-30° C.

DISTINCTIVE CHARACTERISTICS: SL: 200 mm. This form is slightly elongated, i.e., typical of the *Theraps* 2 group. It resembles *C. hartwegi*, but has an irregular pattern of red spots. These red spots are large and cover the entire body, as well as the dorsal fin. The dorsal and caudal fins are edged with bright red and streaked with the same gold color as the body. The black lateral stripe is thin and somewhat broken.

Cichlasoma (Theraps) sp.
"Huitiupan". Rio Colorado,
Mexico.

C. (T.) sp. "Huitiupan"

DISTRIBUTION: On the Atlantic slope of southern Mexico. It is restricted to the upper reaches of the Rio Coatzacoalcos and Rio Grijalva basins.

HABITAT: pH: 7.2-7.5, GH: 2-8, KH: 4-7, T: 23-26° C. This deep–bodied species, oddly enough, is found in the fast flowing waters of the lower mountain regions of southern Tabasco and northern Chiapas. It prefers the deeper holes of these montane rivers, but also lives among the rocks as if it were a slender bodied species.

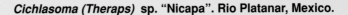

Cichlasoma (Theraps) sp. "Nicapa". Rio Platanar, Mexico.

DISTINCTIVE CHARACTERISTICS: SL: 200 mm. The body is an overall golden yellow, with a bright pink chest, upper belly region, and lower facial area. The center of each scale is pink, and the fins are spotted in pink. A broad, concave band extends from the gill cover to the caudal peduncle.

C. (T.) sp. "Las Flores"

DISTRIBUTION: On the Pacific slope of Guatemala in the rivers that flow into Lago de Guija.

HABITAT: pH: 7.4-7.6, GH: 13-15, KH: 5-6, T: 26-30° C. This species is riverine in nature, occurring in the affluents of Lago de Guija near the Guatemala and El Salvador border.

DISTINCTIVE CHARACTERISTICS: SL: 150 mm. This elongated form is not as slender as its riverine cousins, and its head profile is steeply sloped. Its mouth is small and its lips are rather thin. A broken horizontal stripe extends from an area directly behind the gill cover to a large spot on the caudal peduncle. Another large spot exists on this stripe just posterior to the middle of the body.

The body is an overall olive green. A pinkish red color is visible behind the gill cover and in the shoulder and chest areas. The scales below the mid-lateral line are of this same pinkish color behind the chest region, and turn a deeper reddish or maroon color as they approach the caudal region. Each scale is bordered in the olive color of the body giving it a reticulated pattern. The dorsal and caudal fins are red with sky blue spangling and streaking. The pectoral, pelvic, and anal fins are the same color as the body.

Type species: *Thorichthys ellioti* **Meek, 1904**

The members of this group are found only on the Atlantic slope of Central America from the Rio Antigua drainage of southern Veracruz, Mexico, to the Rio Motogua basin of southeastern Guatemala and northwestern Honduras. These small, slender bodied species have steeply slanted cranial profiles with pronounced snouts and deep preorbital regions. They have small mouths with moderately protractile jaws and eight sensory pores (five on the mandible) between the tip of the chin and the angle of the preopercle. The head and anterior body usually have blue to turquoise spots, and the subopercle usually has a prominent black blotch or spot. Their color pattern is made up of 5-6 vertical bars, or cross bands, behind the head, the third showing a black blotch on or below the lateral line. They have truncate to lunate caudal fins, the lobes of which typically are produced into filaments in the adults, and pectoral fins that extend backward to or beyond the origin of the anal fin. There is an absence of scales at the bases of the soft dorsal and anal fins. Their branchiostegal membranes are hypertrophied, brightly colored, and capable of being distended, forming a conspicuous crest around the throat. This method of behavior is used chiefly by both of the sexes for dominance of territory and courtship. The representatives of this subgenus basically are substratum-sifting invertebrate feeders, much like their large cousins of the *Amphilophus* group. They are capable of reproducing at 8 to 10 months of age

Cichlasoma (Theraps) sp. "Las Flores". Lago de Guija, Guatemala.

and regularly spawn again before their progeny totally disband from the family group, indicating their role as a feeder cichlid highly predated upon.

Cichlasoma (Thorichthys) affine. Lago Peten Itza, Guatemala.

C. (To.) affine (GUENTHER, 1862)

COMMON NAME: Yellow meeki

DISTRIBUTION: Guatemala, Atlantic slope. Restricted to the Lago Peten basin.

HABITAT: pH: 7.4-8.0, GH: 13-15, KH: 5-6, T: 26-30° C. This species is both lacustrine and lower valley riverine, preferring sandy or soft bottoms with small rocks.

DISTINCTIVE CHARACTERISTICS: SL: 140 mm. The juvenile pattern is a silver gray body with 6 vertical bars or cross bands, the middle one of which has a deep black spot where it crosses the mid-lateral line. A small black spot appears most of the time on the bottom of the subopercle. Mature specimens develop a bright yellow ventral area that extends from the caudal peduncle through the throat and onto the face. The pelvic (or ventral) fins and the anal fin have a yellow coloration as well. Bright blue spangles cover the dorsal and caudal fins, and blue streaks outline the rays of the pelvic fins. No spangling is found on the face or the operculum.

C. (To.) aureum (GUENTHER, 1862)

SYNONYM: *Cichlasoma maculipinne* Steindachner, 1864

COMMON NAME: Blue flash

DISTRIBUTION: Guatemala, Atlantic slope. Restricted to the Lago de Izabal basin, including the Rio Motagua and Rio Sarstoon.

HABITAT: pH: 8.0, GH: 6, KH: 4, T: 26-28° C. This species will inhabit warmer lakes and lagoons, but prefers the lower and middle sections of the riverine environment. Some populations penetrate the upper riverine sections, but exist here in limited numbers. The water's high velocity and the lack of adequately sized substrate reduce their ability to proliferate under these conditions. However, they are the second most common *Cichlasoma* subgenus, after *Theraps*, to inhabit these waters.

DISTINCTIVE CHARACTERISTICS: SL: 150 mm. The juvenile color pattern is an olive base with a bronze overtone and 5 vertical bars or cross bands. An interrupted mid-lateral band extends from a spot at the top of the operculum to the base of the caudal fin. A large spot is found just above the lateral band on the middle bar, and a large black subopercular spot is always

Cichlasoma (Thorichthys) aureum. Rio Cienega, Guatemala.

present. Mature specimens develop an iridescent turquoise or blue coloration bordering the bronze colored scales of the body and a bright orange-yellow coloration, with the dorsal and caudal fin a slightly lighter color. The dorsal fin is tipped or edged in red. Turquoise or blue streaks outline the rays of the dorsal, caudal, anal, and pelvic fins, and numerous spots of the same color cover the operculum and facial areas. An iris ring of the same color encircles their unusually large eyes and helps to make this

Cichlasoma (Thorichthys) sp. cf. aureum. Lago de Ilusiones, Mexico.

species a standout. The intensity and extensiveness of the blue coloration vary from one population to another, adding some confusion to its taxonomy.

C. (To.) sp. cf. aureum

COMMON NAME: Gold flash or yellow aureum

DISTRIBUTION: Southeast Mexico, on the Atlantic slope. Extends from the lower and middle sections of the Rio Grijalva system to the headwaters of the Rio Coatzacoalcos system.

HABITAT: pH: 7.0-7.8, GH: 3-9, KH: 3-8, T: 26-30° C.

DISTINCTIVE CHARACTERISTICS: 155 mm. Similar to *C. aureum*, but with a golden yellow coloration or bordering around fluorescent body scales. The spotting of the operculum and face is much more prominent. The body height is greater, the fins are longer, and the eye is slightly larger than in *aureum*. Also, this species develops extremely ling filaments on the caudal fin, the longest of any of the Mesoamerican species.

C. (To.) callolepis (REGAN, 1904)

DISTRIBUTION: Mexico, Atlantic slope. Restricted to the upper tributaries of the Rio Coatzacoalcos, Oaxaca.

HABITAT: pH: 7.2-7.5, GH: 2-5, KH: 4-6, T: 24-26° C. Rocky substrate of the mountain headwaters.

DISTINCTIVE CHARACTERISTICS: SL: 140 mm. This species is

Cichlasoma (Thorichthys) callolepis. **Illustration by John R. Quinn.**

the most slender bodied of this subgenus, allowing it to inhabit the faster flowing waters of the upper sections of the rivers or headwaters. Its eye size is the largest of this complex, but the preorbital is not very deep. There is only a slight trace of any subopercular black spot. Deep orange spots arranged in horizontal rows extend from the operculum to the base of the caudal fin. The throat is reddish-orange, marked in wavy streaks. The anal and pelvic fins, and the base of the

caudal fin, are also orange. The dorsal fin below the white band shows a hint of orange. The anal fin has pale blue spots, and pale blue spots occur in a row following the curvature of the orbit, directly under the eye and on the preopercle.

C. (To.) ellioti MEEK, 1904

DISTRIBUTION: Eastern Mexico, Atlantic slope, from the Rio Papaloapan system to the Rio Coatzacoalco system.

HABITAT: pH: 7.0-7.5, GH: 6-8, KH: 8-10, T: 25-28° C. This species populates the lower

Cichlasoma (Thorichthys) ellioti. **Rio Naranja, Mexico.**

and middle river valleys over sandy and rocky substrate.

DISTINCTIVE CHARACTERISTICS: SL: 150 mm. This species seems to be the link between *C. meeki* and *C. helleri*, resembling both to a degree. Juveniles show a prominent 6 bar pattern on a silver gray body, with a large spot on the middle bar crossing the mid-lateral line. They are deep bodied at the forward point of the dorsal, but the body slants drastically so that it is rather thin and shallow bodied at the rear point of the dorsal. Adults display a bright red coloration on the belly and lower half of the operculum. A bold, black blotch appears on the subopercle and is encircled in pale blue. Its pelvic and anal fins are a bright orange-yellow and streaked in pale blue. The

dorsal and caudal fins are streaked in this blue coloration, but have no spots or specks. A row of small blue spots follow the curvature below the eye and a few appear on the face and lower mandible.

C. (To.) helleri (STEINDACHNER, 1864)

DISTRIBUTION: Mexico, Atlantic slope. Restricted to the lower and middle river valleys of the Rio Grijalva system. Rio Mezcalapa to the Rio Tacotalpa.

HABITAT: pH: 7.2-7.5, GH: 2-8, KH: 4-7, T: 23-26° C.

DISTINCTIVE CHARACTERISTICS: SL: 145 mm. Juveniles have a light violet body coloration with 5 vertical bands. Adults show a pinkish-violet color with 7-8

Cichlasoma (Thorichthys) helleri. **Rio Tacotalpa, Mexico.**

horizontal rows of turquoise colored spangles through the middle of the body. Its dorsal and anal fins are a light orange coloration and spangled with small turquoise colored spots. The pelvic and caudal fins are streaked and spangled. Two rows of small spots follow the curvature below the eye, and a few larger spots occur just above and behind the eye.

C. (To.) meeki (BRIND, 1918)

DISTRIBUTION: Mexico, Guatemala, and Belize, on the Atlantic slope. Ranges from the Rio Mezcalapa, a tributary of the lower Rio Grijalva, to the

Cichlasoma (Thorichthys) pasionis. Rio de la Pasion, Guatemala.

Rio Sarstoon.

HABITAT: pH: 7.4-8.0, GH: 13-15, KH: 5-6, T: 26-30° C. Widespread throughout the Rio Usumacinta basin, but is not found in the Peten basin. This rather robust species prefers the lower and middle sections of the rivers in the slower moving waters. It prefers a soft substrate of sand and mud. It commonly is found in the floodplains created by heavy rainfall.

DISTINCTIVE CHARACTERISTICS: SL: 155 mm. As a juvenile, it has a silver-blue body coloration with a mid-lateral band broken by a black blotch in the middle of the body. Vertical bars are rarely present, including the fry stage and fright pattern. A scarlet red face, throat, and abdomen develop at an early stage in its life. The scales below the lateral line are colored blue, but the area above the lateral line is silver-gray. Blue spangling is present in the dorsal, caudal, and anal fins. The pelvic fins are streaked rather than spangled.

C. champotonis and *C. hyorhynchum* may be junior synonyms, but further work must be done on these species.

C. (To.) pasionis RIVAS, 1962

DISTRIBUTION: Mexico and Guatemala on the Atlantic slope. Widespread, ranges throughout the Rio Usumacinta system from the Rio Mezcalapa in SE Mexico to the Riachuelo Machaquila in Guatemala.

HABITAT: pH: 7.4-8.0, GH: 13-15, KH: 5-6, T: 26-30° C. Widespread throughout the Rio Usumacinta basin, but is not found in Lago Peten. It is both riverine and lacustrine and prefers the softer substrate.

DISTINCTIVE CHARACTERISTICS: SL: 170 mm. This is the most deep bodied and robust of this subgenus and, like *C. meeki*, it is widespread throughout the Rio Usumacinta basin and portions of the lower Grijalva system. Both species are replaced in Lago Peten by *C. affine*, but *C. meeki* inhabits the river systems to the east in Belize, while *C. pasionis* populates the river systems to the west and south. Juveniles have a violet-gray body coloration, with a broken or interrupted lateral line broken by the mid-lateral black blotch. A bold, black, subopercular spot is prominent, and even the lower curvature of the subopercle is edged in black. Adults develop a bright yellow coloration over most of the body, with incomplete horizontal rows of green colored scales that extend from the abdomen to the base of the anal and caudal fins. Two to three rows continue above the mid-lateral line and extend to the base of the caudal and dorsal fins. The anal, caudal, and dorsal fins are spangled in turquoise, and the pelvic fins are streaked in turquoise.

C. (To.) socolofi MILLER & TAYLOR, 1984

DISTRIBUTION: Southeast Mexico, Atlantic slope. Found in the tributaries of the upper reaches of the Rio Tulija of the Rio Usumacinta system. Rio Mi-sol-ha.

HABITAT: pH: 7.5-8.0, GH: 13-18, KH: 14-18, T: 23-25° C. This species prefers the faster flowing waters of the middle and upper reaches of the riverine environment.

DISTINCTIVE CHARACTERISTICS: SL: 140 mm. This slender bodied species is the most elongated of this subgenus, similar to *C. callolepis*. Its mid-lateral band is extremely wide and the mid-lateral blotch is broad and tall. The body coloration is bronze, with a bright pink face, belly, and abdomen, the color extending to the caudal peduncle. Several horizontal rows of blue-green spangles extend from the operculum through the caudal peduncle to the caudal fin. These rows are placed from the belly upward to above the mid-lateral line, but do not cover its back. The face has very few spots, if any, and the operculum is streaked in blue, as are the pelvic and anal fins. The caudal is almost transparent, but the dorsal fin is streaked in a pastel orange between the hard rays.

Cichlasoma (Thorichthys) meeki guarding young. Photo by Hans-Joachim Richter.

Other North & Central American Cichlids

AEQUIDENS

A. coeruleopunctatus (KNER & STEINDACHNER, 1863)

DISTRIBUTION: Southern Costa Rica and Panama, on the Pacific slope.

HABITAT: pH: 6.7-8.2, GH: 3-6, KH: 3-6, T: 22-29° C. It lives in stagnant waters as well as in the shallows of smaller rivers with organic material and low velocity. They thrive on aquatic insects.

DISTINCTIVE CHARACTERISTICS: SL: 145 mm. This species is the only representative of its genus in Central America. It has three anal spines, the least of any cichlid in Central America. The body is very

Aequidens coeruleopunc tatus. Rio Coloradito, Panama.

robust, and its head profile is almost flat. The body is emerald green. They school together in small groups on the river bottoms.

GEOPHAGUS

G. crassilabris STEINDACHNER, 1877

DISTRIBUTION: Panama, on the Atlantic and Pacific slopes.

HABITAT: pH: 7.0-7.9, GH: 3-4, KH: 3-4, T: 22-30° C. This species prefers moderate to fast flowing water over a substrate of small stones, sand, rocks, and leaf debris from the forest canopy.

DISTINCTIVE CHARACTERISTICS: SL: 240 mm. Juveniles of this elongated species have a gray coloration, with an interrupted lateral band and 6 vertical bars. Adults become a rich bronze color and the vertical bars turn a green color. Turquoise colored scales cover the majority of the body and an iridescent turquoise patch is evident on the operculum. The fins are all edged in this turquoise coloration. The upper lip extends considerably beyond the lower lip and the head develops more of a knot than a nuchal hump.

HEROTILAPIA

H. multispinosa (GUENTHER, 1866)

NATIVE NAME: *zarzapala*
COMMON NAME: Butterfly cichlid

DISTRIBUTION: Honduras and Costa Rica, both the Atlantic and Pacific slopes. Ranges from the Rio Patuca in Honduras to the Rio Matina in Costa Rica.

HABITAT: pH: 6.3-7.2, GH: 2-4, KH: 2-5, T: 21-36° C. This species finds refuge in lakes and swampy areas with muddy bottoms. It forages in the mud for anything edible and eats filaments of algae.

DISTINCTIVE CHARACTERISTICS: SL: 90 mm. This species is quite small. It has a golden color, orange eye, a black band between the eyes, and a spot situated at about the halfway point of the body. It has 7-8 wide dark bars in the posterior half of the flank. The dorsal fin has blue tones, the anal and pelvic fins are dark, and the pectoral fins are light. It has tricuspid teeth, a unique characteristic of the Central American cichlids.

NEETROPLUS

N. nematopus GUENTHER, 1866

COMMON NAME: Poor man's tropheus

DISTRIBUTION: Nicaragua and Costa Rica, Atlantic slope. This species resides in the Great Lakes of Nicaragua and their affluents, and the tributaries of the Rio San Juan.

HABITAT: pH: 7.0-8.75, GH: 3-18, KH: 3-21, T: 21-34° C. This species is both lacustrine and riverine. It is found in all rivers, but proliferates in rivers with medium to high velocities. They live upon the *aufwuchs* that cover the rocks and submersed trees and filaments of algae.

DISTINCTIVE CHARACTERISTICS: SL: 140 mm. This species is small. It is easily recognized by its curved snout, inferior mouth, and elongated body. A

Geophagus crassilabris. Rio Chagres, Panama.

***Herotilapia multispinosa*. Rio Zapote, Costa Rica.**

wide dark bar crosses the middle point of the body. Their teeth are truncated in the form of incisors. Their general body coloration is gray, darker dorsally and lighter ventrally. Approximately seven dark bars are evident in the juvenile pattern, but are almost absent in adults, with the exception of the third bar which remains in all ages. The iris is silver and frequently bluish. The unpaired fins are dark brown and the pectoral fins are light. The pelvic fins are dark, except for the first soft ray which is white. During the periods of high temperatures the body becomes black and the third bar becomes white, resembling *Tropheus duboisi* from Lake Tanganyika in Africa.

PARANEETROPLUS

This genus differs from *C. (Herichthys)* in having broader and more strongly compressed teeth. The outer series in the lower jaw is transverse, formed of subequal teeth with rounded or obtusely pointed apices.

P. bulleri (REGAN 1905)

DISTRIBUTION: Southern Mexico, Atlantic slope. Endemic to the upper reaches of the Rio Coatzocoalcos basin in Mexico, in the rivers Rio Almoloya, Rio Jaltepec, Rio Malotengo, and Rio Sarabia in the state of Oaxaca.

HABITAT: pH: 7.2-7.5, GH: 2-5, KH: 4-6, T: 23-25° C. It prefers the fast flowing water of the montane environment. This species consumes heavy amounts of algae and eats from the *aufwuchs* that cover the underwater biotope.

DISTINCTIVE CHARACTERISTICS: SL: 255 mm. This slender bodied species is quite large as compared with similar *Theraps* species. Its mouth is placed on the ventral side of its sharply convex head, like the *Labeotropheus* species from Lake Malawi in Africa. Juveniles have the basic gray coloration, with 6 black blotches along the longitudinal line. Adults exhibit a bright yellow coloration above the lateral band and a yellow face and head. The dorsal, anal, pelvic, and pectoral fins also are yellow. The ventral area beneath the mid-body line is white, as is the caudal fin. A rosy pink patch covers the operculum and the area just above and behind it. In some specimens this patch extends into and across the forehead.

PETENIA

The single species is related to *C. dovii* and *C. managuense*, but has the mouth larger and the maxillary more exposed.

P. splendida GUENTHER, 1862

DISTRIBUTION: Mexico, Guatemala, and Belize, on the Atlantic slope. Found from the Rio Papaloapan basin and throughout the Rio Usumacinta system to the Belize River.

HABITAT: pH: 7.4-8.0, GH: 13-15, KH: 5-6, T: 26-30° C. Widespread throughout the Rio Usumacinta basin, and is found also in the Lago Peten basin. This rather robust species prefers the lower and middle sections of the rivers in the slower moving waters. It prefers a soft substrate of sand and mud.

DISTINCTIVE CHARACTERISTICS: SL: 50 cm. This species is lacustrine in nature, but also inhabits the lower river valleys. Being piscivorous, this species

***Neetroplus nematopus*. Lago de Nicaragua, Nicaragua.**

is at the top of the trophic pyramid and, therefore, is not abundant. They always are on the hunt for smaller fishes on which to feed. Their torpedo body shape allows them to dart through the water with great bursts of speed, enabling them to suck in their quarry with their highly protrusile jaws. Their pseudocanines can hold firmly onto bigger prey that are too large to swallow. The body coloration is a brown-gold.

***Paraneetroplus bulleri*. Rio Jaltepec, Mexico.**